SELF-FAB HOUSE

2nd Advanced Architecture Contest

organized by

Iaac
Institute for
advanced
architecture
of Catalonia

in collaboration with

MINISTERIO
DE VIVIENDA

fundació caixa d'arquitectes

ACTAR

DO IT YOURSELF

AUTO- & BIOGENERATED

MATERIALS & CONSTITUTION

RECYCLED & READAPTED

MODULATED & PRE-ARRANGED

DIGITAL FAB

INTRODUCTION
by Beatriz Corredor · Minister of Housing

Housing that is accessible to all citizens, which is a constitutional right and one of the duties of every public administration, must also strive for the highest functional, material and architectural quality. New technologies make it possible to achieve greater quality in house production, with a fuller involvement at the local level.

The Self-Fab House competition organized by the Institute for Advanced Architecture of Catalonia, for proposals for new ways of making homes using local technology, achieved the participation of over 500 architects and architecture students from 87 countries who contributed ideas and solutions for building homes more efficiently, using cleaner technologies, and reducing the impact of the construction process on the global environment.

The active involvement of local actors and the use of locally available resources favour a process that is fundamental to ensuring that housing has not only a quantitative but also a qualitative value, in which the home, the first basis for future projects in life, is constructed in an open and transparent process that creates the highest quality at the lowest price, stimulating the socialization of the habitat and its integration into the area's economic, social and cultural structures.

It is here that platforms for research and debate such as the Institute for Advanced Architecture of Catalonia are endeavouring to go beyond the limits of informed discussion and have an effective impact on a major industry, the construction of homes, and also on the construction of the city, defined by the new habitats in which we can develop our potential to the full. In this endeavour they will always have the support of the Ministry of Housing.

FOREWORD

by Francesc Fernandez · President of IaaC

The Institute for Advanced Architecture of Catalonia has held the second Advanced Architecture Contest: THE SELF-FAB HOUSE, for the purpose of stimulating new research into housing. The success of the first competition, which attracted entries from more than 2,000 architects and architecture students from 108 countries, has prompted the IaaC to organize this second competition as a platform for innovation in the physical production of the conditions of human habitability. The IaaC is an educational and research centre based in Barcelona, an international acclaimed leader in its field, which welcomes students from over 25 countries, so that any meeting or discussion that takes place there benefits from the inputs of architects from many different cultures and backgrounds. This diversity is a key characteristic of contemporary society, and one we must incorporate into our reflections on urban culture and architecture in the interests of hybridizing processes and sharing ideas and solutions between countries.

In fact, the architecture and the construction of housing is undergoing a process of transformation as it embraces other issues relative to energy and the management of resources that were not formerly part of the construction industry. Building homes and the city today also means identifying and preparing where the resources they consume will be produced. This means that the production of housing does not end when a building is completed, because just as important as constructing a building well is knowing how to manage its running over time. From the IaaC we want to congratulate the winners of the competition and everyone who took part and encourage them to ensure that the ideas presented here help transform their communities and cities.

THE SELF-FAB HOUSE
by Vicente Guallart · Director of IaaC

Housing is the direct expression of the society that inhabits it. The first vital need that led to the emergence of architecture was the need for shelter in a dynamic medium whose atmospheric conditions changed in the courses of the day and the year. The home is the largest individual skin, marking the boundary between public and private. And what was originally a matter of transforming a natural material — stone, wood, straw — from the immediate environment was itself transformed by the growth of human communities in cities into a more complex process involving construction systems based on grouping and superposing, in which the production of housing was no longer tied to its user but entailed a much larger economic process.

In recent years, housing in industrialized countries has lost much of its intrinsic value, becoming just another product of the consumer society, with the price of the plot far exceeding the material value of the actual building.

At the same time, the processes of social and technological participation in our social networks or the free code with which part of the Internet works have not yet been translated into new ways of constructing the spaces in which we interact with these networks. Is the physical space merely an amorphous container devoid of cultural and emotional intensity, or can it potentially transform into an infrastructure that literally emerges from the processes and systems of the Internet? If the homes of the Machine Age eventually arrived at more flexible and open structures capable of incorporating technologies that ensured a certain degree of comfort based on the consumption of resources produced in remote locations far from the cities, how can we construct homes and habitable structures consistent with the networked world we live in?

The Institute of Advanced Architecture of Catalonia is an active partner of The Center for Bits and Atoms at MIT in the setting up of Fab Lab fabrication laboratories. Anyone can produce anything anywhere in the world if they have the requisite knowledge and machinery. Instead of a world of consumers of resources and information, the Internet Age is creating a society of producers in which everyone has the potential to produce energy, food and commodities on the basis of networks of shared knowledge.

We are committed to the idea that housing can be self-fabricated using knowledge and processes that are shared through networks and local resources and machinery, making the production of a home specific rather than generic, a process that responds to the particular environmental conditions and mainly uses materials from local sources. And to the extent that the developer is no more than an intermediary between people needing homes and the banks, that role may be greatly reduced or even disappear. Contributing added value and developing processes that stimulate the future occupants will be key functions of those who want to lead the processes of housing production.

ARCHITECTURE IN OUR HANDS

by Lucas Cappelli, Director Advanced Architecture Contest

If a system were implemented to *share architectures* for the local construction of houses using new technologies, we might suddenly find an individual downloading a free program that enabled them to manufacture and construct a house, with predetermined parts made of local materials, adapting it to their own requirements and personalising it right down to the last detail.

The idea of solving such a widespread basic social problem as housing using global knowledge, shared experiences and the resolution of similar problems in very different places might well be addressed by harnessing the new communication and technology systems in conjunction with the notion of personal manufacturing.

At a glance, it certainly appears that self-sufficient models of this kind could well be the long-awaited solution to what we architects have been mulling over for some time: a model enabling the ego to be excluded as an end in our discipline, to finally escape from the narcissistic circle in which we have entrapped ourselves in recent years, giving free rein to our boundless vanity.

The self-fab house is, in essence, merely an excuse; it is yet another martyr to the generation of knowledge. It represents generosity, the suicidal capitulation of the architect, our infinite surrender, the understanding that for architecture to serve its deepest intention, the star architect has to disappear; the superfluous, the artificial, the annihilating effect of the superimposed, of the infamous, the unnecessary and the pornographic exposition of egotistic desires has to die. Only Architecture should remain.

The modernists had the courage to mention this. Houses are habitation machines. Buildings should represent their users in every dimension, be their reflection or counter-reflection, their truths and their lies, but it is the inhabitants themselves that need to reflect all of this.

We no longer want to talk about paradigms in architecture; it is now, still without credentials, that we step across the border from the real into the whole world, where space might receive its deserved extent, its perfect mould, and its long-awaited and much-talked-of response.

Our soul mate could be here; we could inhabit it, stretch out our arm and open a door. It is now that my window takes the form of my ears because I prefer to hear and not see. It is a historic moment; it is time to make sacrifices and to make room for the architecture of the people. And to become true architects.

MY HOUSE

by Willy Müller · Co-Director Master in Advanced Architecture

When we talk about self-fabricated houses we are probably talking about most of the domestic architecture in the world.

This is a long and profound global experience on which to capitalize. The output of architecture d'auteur in the best of cases, or of engineering firms large and small in the worst, with a whole host of pseudo-experimental institutional production in the middle, has sufficient media power virtually to eradicate from the iconic references of the house this common historical, cultural and partly atavistic practice of construction: that of 'my house', that form of moving sideways through architecture without mediatizing the process between the idea, the need and the available potential.

At the present time self-constructing is interesting precisely on account of this lapse. The loss of inherited practices, the historical amnesia regarding the need to build my own habitat in the most developed settings, as if there had always been municipal housing institutes, architecture firms in the Yellow Pages and a branch of Habitat or similar everywhere, all over the world, makes the intention original again, recovering the virginity of thinking about building 'my house'.

What does self-constructing mean today? It is clear that whether we look at this transverse practice, common to all cultures and geographies, with the attitude of the most closed and isolationist tradition of a Mongolian village or with that of the uprooted inhabitants of any Latin American city, displaced and deprived of identity, well-being and access, we find not only the most varied and original ways of thinking about 'my house' but also a number of common features, such as the saving of energy and the eradication of waste, the exaggeration of utility and the contradiction between being part of a camouflaged community, and therefore safe from danger, and the naïf superposing of an element of original identity converted into a message: 'this is my home'.

Nowadays we have sophisticated tools for extending this attitude, but uprooted from tradition, delocalized of marginality and mediatized by the work of an architect who signs the process, explains and disseminates it. This is a big change, with inevitable consequences, and one that forces us to consider the benefit in quotas of sustainability and assume the risk of self-constructing not out of obligation or necessity but as an act of commitment or responsibility.

Now that we architects have got to this point in awareness, preparation and tools, let us not forget that other powerful productive sectors such as the car industry have already managed to put mass-produced homes on the market, with sufficient standards of design, comfort and energy saving to enable a leap in production at some point in the decline of the motor car as we undertake a change of system and move from the age of oil to the age of hydrogen.

This is a very old way of thinking, but with innovative results in the here and now that can be confusing. This form of production implies huge factories, serialized designs, stocks at the global scale and costly logistics, though the value of the product is continually falling.

It is thinking 'my house' many times, when the really big change lies in thinking 'my house' just once.

But it will mean a radical change. If we can think of self-constructing, and do so in a unique, personal and decentralized way, we will enter a world of concepts that we must address, in addition to those of non-serial production:

Can we self-construct complex multi-functional buildings, or can we only bring about change in the smallest and most mobile?

Will we have to accept a greater consumption of land as a result of these practices of self-construction, or do we have new ideas about how to make houses that can be swapped around, stacked up or slotted together?

Must we reinvent density urbanism in order to make self-fabrication possible?

Do we think in terms of plots or party walls or only of the bounds of a single space, moving from an m^2 urbanism to an m^3 urbanism?

Do we go back to constructing in situ, as a return to a past 'made with my own hands', taking the machinery to the site, or do we delocalize construction and assembly, without the logistics of mobility leading to a loss of energy efficiency?

Can we think of houses to bring to 'my house'?

Responding to these challenges is always a case of asking new questions.

Or rather, of asking once again the question: 'How will I make my house?'

This is what we are doing at IaaC.

Petrocelli was a legal drama which ran on NBC from September 11, 1974 to March 3, 1976.

It had 48 episodes.

Tony Petrocelli was an Italian-American Harvard-educated lawyer grew up in South Boston who gave up the big money and frenetic pace of major-metropolitan life to practice in a sleepy city in the American Southwest called San Remo (filmed in Tucson, Arizona). He and his wife Maggie lived in a trailer in the country while waiting for their new house to be built, and travel around in a beat-up old pickup truck.

OPEN SELF-STANDARDS

by Marta Malé-Alemany · Co-Director Master in Advanced Architecture

Despite their late appearance in the field of architecture, CAD-CAM technologies and digital numerical control manufacturing tools are being adopted in many industrial sectors for the purpose of design and construction. An increasing number of industrialists own or have access to this type of machinery, which means that the architect's sphere of activity is full of new opportunities. Beyond the private industry of developed countries, schools of architecture have begun to realise that these tools are essential for turning virtual, three-dimensional models into physical and material constructions. For the contemporary architect, being able to create a material prototype on their own computer is invaluable: it not only connects them to the industry that will carry out their work, but also opens up new forms of exploration that had been eliminated by the industrialisation of the 20th century.

From this perspective of research and the use of new technologies, IaaC created a digital manufacturing centre (Fab-Lab) that is now equipped with several laser cutting machines, milling machines and 3D printers for students' and professionals' use. The Fab-Lab is a place where everyone can personalise and manufacture their own life-size designs in a range of materials that include plastic, wood and metal. In keeping with the philosophy of bringing self-fab closer to the people, Fab-Lab belongs to a group of centres distributed throughout the world (with which we connect and exchange knowledge in real time via video conferencing), and which, in turn, promote the creation of other Fab-Labs around the globe, in much less privileged areas. This idea and commitment, which breaks away from traditional cooperation models, is based on a simple principle: we can do a lot more by creating a Fab-Lab and sharing our knowledge than by simply sending prefabricated parts to these places. We give them the opportunity to imagine and construct their own tools, appliances, furniture and —why not- dwelling places, locally.

The SELF-FAB House competition was designed so that everyone would have the opportunity to manufacture their own design, directly or indirectly, through centres like the Fab-Lab, or others of a similar nature that were open and shared. With our eyes set on this fast approaching future, we wanted to receive self-fab proposals from all over the world; proposals that were not restricted by predetermined scales or programs on inhabitable spaces, and where the architect was at complete liberty to choose the construction materials and project and/or manufacturing technologies. We at the IaaC wanted the competition to provide an environment for the exchange of knowledge and work methods, combining traditional processes with the latest cutting-edge techniques. Fuelled by the energy and positive results of our first competition on SELF-SUFFICIENCY (the construction of self-sufficient buildings), we believed it was important to extend the idea of self-sufficiency to each person's capacity to build their own habitat using local resources and global technology. Thus, the competition encouraged participants to go beyond the limits of standard solutions and to explore more sustainable decentralised production models. We wanted to see how we could continue learning from local techniques and processes, while using technology to explore a constructive and formal field of experimentation that does not exclude housing.

CAN WE RECYCL
IN ORDER TO EM
OUR SELF-FABRIC

E USED OBJECTS
BED THEM INTO
ATED DWELLING?

CAN WE SELF-F
OWN DWELLING
WE DO WITH OU

ABRICATE OUR
THE SAME WAY
R FURNITURE?

IS ADVANCED TE
FERING TO MAS
TION OR TO TH

CHNOLOGY RE-
SIVE PRODUC-
E LOCAL ONE?

CAN WE COMBINE TRAD
ION VALUES OF SELF-S
CONTEMPORARY TECHN

ITIONAL CONSTRUCT-
FFICIENCY WITH THE
OLOGICAL ADVANCES?

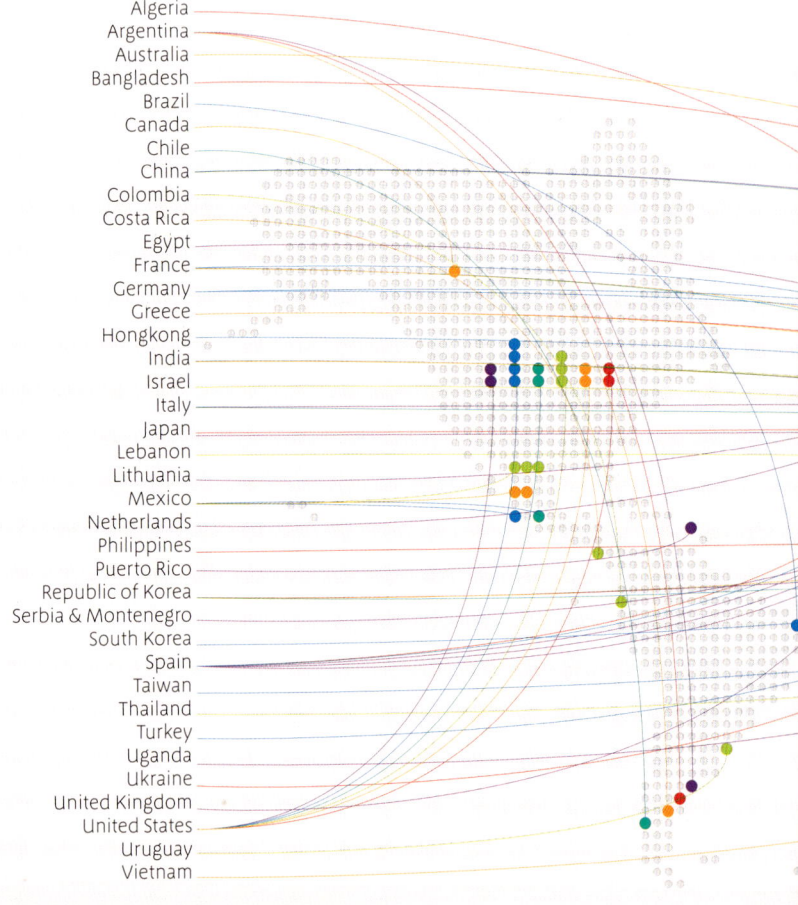

Algeria
Argentina
Australia
Bangladesh
Brazil
Canada
Chile
China
Colombia
Costa Rica
Egypt
France
Germany
Greece
Hongkong
India
Israel
Italy
Japan
Lebanon
Lithuania
Mexico
Netherlands
Philippines
Puerto Rico
Republic of Korea
Serbia & Montenegro
South Korea
Spain
Taiwan
Thailand
Turkey
Uganda
Ukraine
United Kingdom
United States
Uruguay
Vietnam

IAAC JURY

Vicente Guallart Director IaaC
Lucas Cappelli Director Advanced Architecture Contest
Willy Müller Co-Director Master in Advanced Architecture
Marta Malé Alemany Co-Director Master in Advanced Architecture
Rodrigo Rubio Winner 1st AAC
Daniel Ibañez Winner 1st AAC

GUEST JURY

China · Yung Ho Chang
Head of the Department of Architecture, MIT

Japan · Turlif Vilbrandt
Director of Technology, Digital Materialization Group

Korea · Young Joon Kim
YO2 Architects (Seoul)

Mexico · Michel Rojkind
Rojkind Arquitectos

Spain · Josep Lluís Mateo
Map Arquitectos

Taiwan · J.M. Lin
The Observer Design Group (Taipei)

Uruguay · Julio Gaeta
Director El Arqa

USA · Greg Lynn
Institute of Architecture die Angewandte

Firstly, the jury would like to thank all of the architects and students who took part in the competition, bringing very diverse points of view and cultural conditions to bear on their ideas of self-construction in order to foment and develop new construction systems around the world.

1st Prize
China · Ming Tang
Dihua Yang

This proposal uses a traditional local material and successfully implements geometric elements to create self-transforming and re-informing structures. The jury was impressed by the integration into the landscape and the possibility of being constructed as prototype.

2nd Prize

Spain · Luis Aguirre Manso

The jury was captivated by the hybridization of lightweight construction systems that rise from the ground, and the functional scheme surrounding the chimney, which draws on the principles of traditional architecture.

3rd Prize

Hong Kong · Shinya Okuda, Kung Yick Ho Alvin, Lam Yan Yu Ian

The jury was fascinated by the use of advanced technologies in the manipulation of biodegradable materials to create a system that can be assembled as a sustainable construction.

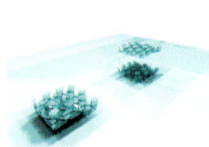 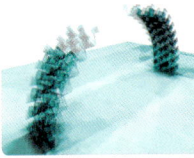 

Mentions The jury is also pleased to give two honorable mentions.

Mention 1

France · Queney Sebastien, Chevance Sebastien

The jury was very satisfied by the possibility of a house that can be constructed from a single material using digital fabrication processes.

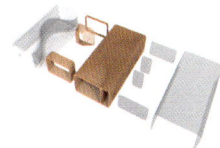

Mention 2

Romania · Barsan Pipu Claudiu
Dragan Andrei
Nituica Oana Maria

The jury was impressed by the social dimension of the proposal in exploring the nomadic approach to construction, based on very lightweight construction systems.

DO IT YOURSELF
DIY

This category is trying to point out the simplicity of the construction process as long as the variations of the final products that could be developed, once a dwelling is assembled by its own user.

Moreover, is promoting the notion of the customization of production as long as the personal expression by modifying the building systems during the construction; not only depending on the way a house is built but also on the materials that are finally chosen to be used.

The essence of Self-Fabrication and Self-Construction has been intended, in some cases, by the establishment of new "smart-shaped" modules that can be put intelligently together. In this case, assembling of your own home could be as simple as a kid's toy. The relocation of each piece could result in both different shapes and spatial variations.

Additionally, there have been proposed the re-use, recycling or reconfiguration of older materials or even objects. There have been examined the possibilities of reusing older building parts left after a severe destruction, or even

DO IT YOURSELF

AUTO- & BIOGENERATED

MATERIALS & CONSTITUTION

RECYCLED & READAPTED

MODULATED & PRE-ARRANGED

DIGITAL FAB

the recycling of some objects that seem to have lost the characteristics of their older configuration, though they could be very useful as building modules. In both cases, the "do it yourself " notion is highly encouraged.

The building systems proposed have the ability to both be assembled and disassembled. This fact is considered to be a crucial one, considering that the materials, objects or modules used could be themselves recycled for a series of times. Moreover, these building systems enable the possibility of expansion of the structure very easily, by only adding the established modules or parts.

Eventually, the non-necessarily specialized labour, as long as their ability to be immediately applied in case of emergency are considered very important. The creation of such a system of self-fabrication and construction of one's own dwelling would result in the decentralization of the building production and industry. The localization of production would have a severe effect on the contemporary functional relationships of building production, by making them more sustainable and less time-consuming.

78FDOD

RIVAS RUZAFA, Elena

Escuela Técnica Superior de Arquitectura
de Madrid (ETSAM)

erruzafa@hotmail.com

In this project, self-construction is thought of as one of the production systems that underpin the contemporary way of life. It is thus regarded as essential to make effective use of this system of production, together with its social, political, technological and aesthetic qualities.

- Sophisticated systems v. traditional systems: High-technology materials v. conventional materials. The project proposes the use (or re-use) of materials with a totally new perspective. They can be adapted, transformed, incorporated into a new cultural use or re-use, construction and deconstruction, expiration and time-cycles.

- Self-construction of territory and field of actuation: We self-construct our own field of work, we have certain tools and we not only choose the territory in which we investigate, but also the objects and agents involved.

- Squamous construction of a tissue: Accommodation should be available during all phases of construction. Each user is the owner of each independent piece. The tissue will be constructed in a collective fashion and according to the agreements established. The tissue also functions as a support structure for services, formwork and bearing structure. The materials and the construction tools used are essential to the final product and reflect the economic, social and technological conditions obtaining during the period of construction. The aim of the system is to construct simultaneity between different technologies, which can assume the global (external) and internal changes of the process. The Layers established adapt to different seasons and geographical situations to ensure comfort.

- Pipe Network: There is an independent construction for each element of the overall structural system, or in some cases a parallel construction that promotes the overall development. It is built up by aggregation.

- Users and Association systems: There is a perceived need to set up associations. There is no such thing as absolute self-sufficiency. Every process results from the associations of different qualities that define the final project. From acquiring the land to the sale of individual pieces.

- Freight and management systems, Costs and expiry, Weather, Plant Ecosystems also figure among the parameters permitting the development of a complex project that will never be finished.

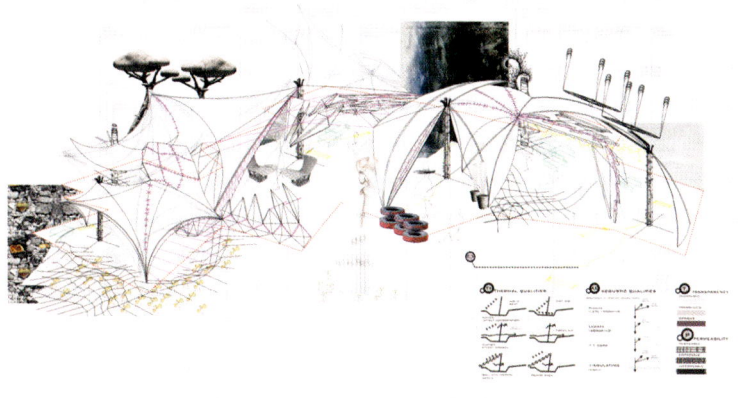

DO IT YOURSELF

AUTO- & BIOGENERATED

MATERIALS & CONSTITUTION

RECYCLED & READAPTED

MODULATED & PRE-ARRANGED

DIGITAL FAB

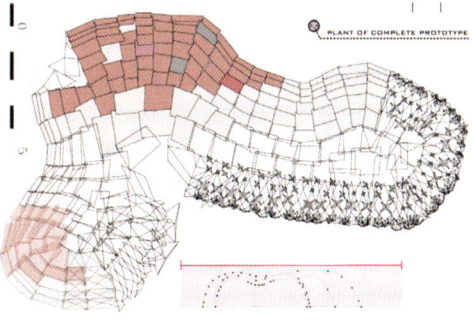

PLANT OF COMPLETE PROTOTYPE

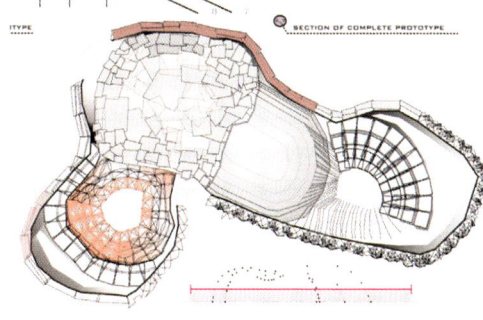

ITYPE

SECTION OF COMPLETE PROTOTYPE

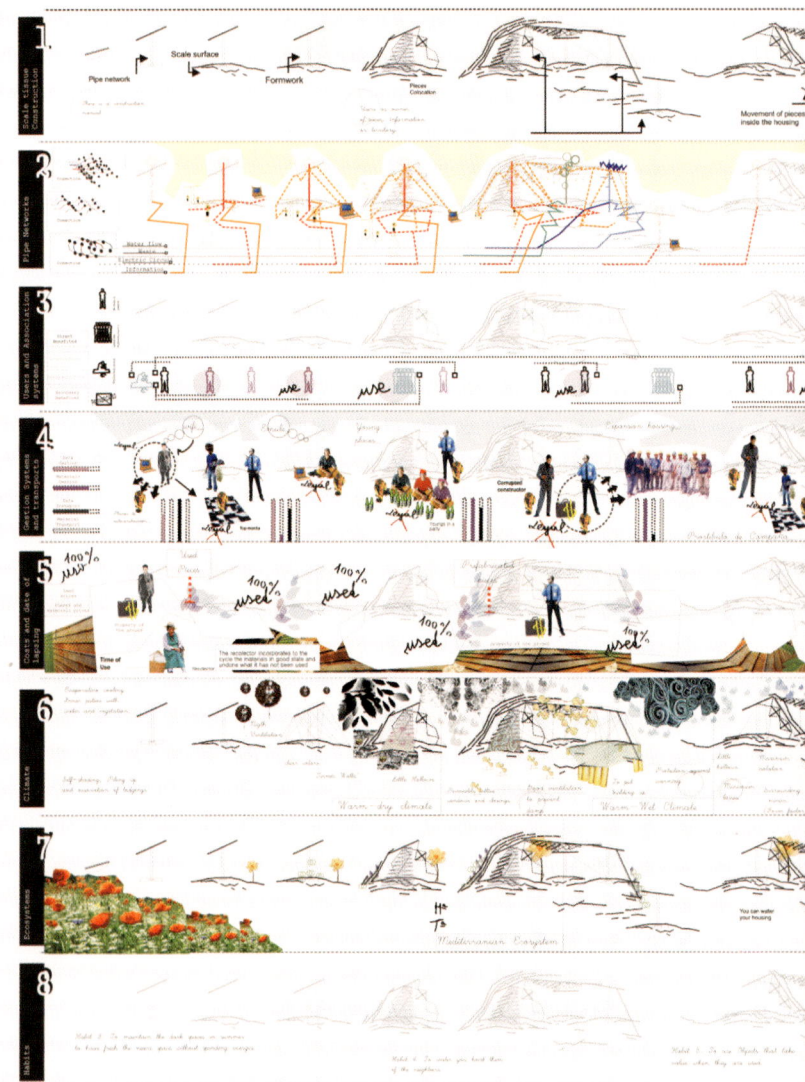

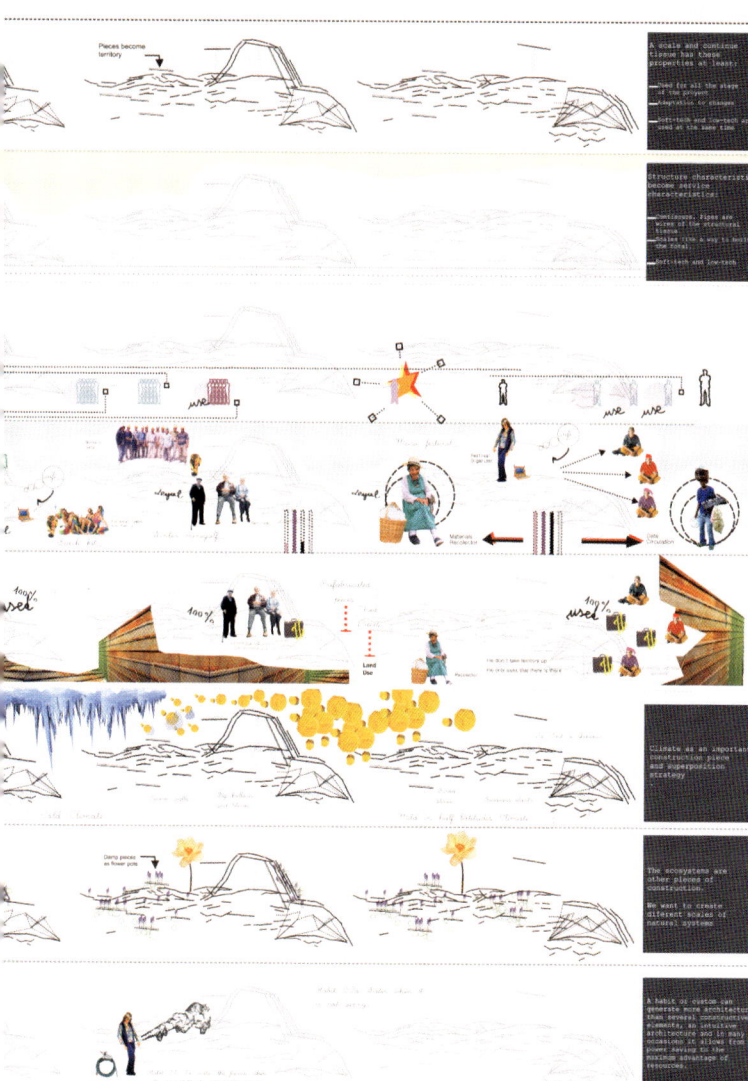

DO IT YOURSELF

AUTO- & BIOGENERATED

MATERIALS & CONSTITUTION

RECYCLED & READAPTED

MODULATED & PRE-ARRANGED

DIGITAL FAB

A scale and continue tissue has these properties at least:

- used for all the steps of the project
- integration of changes
- soft-tech and low-tech are used at the same time

Structure characteristics become service characteristics:

- Continuous. Pipes are wires of the structural tissue.
- Scales (the a way is built the form).
- Soft-tech and low-tech

Climate as an important construction piece and superposition strategy

The ecosystems are other pieces of construction.

We want to create different scales of natural systems.

A habit or custom can generate more architecture than several constructive elements, an intensive architecture and in many occasions it allows free power saving to the maximum advantage of resources.

CAUMERON, Jimmy
GRANADOS, Tristan
REYES, Camille

STUDIO FLAG

jimcaumeron@yahoo.com

For the purpose of providing cheap sustainable housing, the Bayanihan movement is a logical universal mode of instruction based on human proportion. The objective is to decentralize the production process in factories and shift it closer to the user. This minimizes the problems of logistics and transportation as well as bottlenecks in the production process, while allowing limitless possibilities in the spreading of concise general information by way of snail mail, email, inclusion in product packaging, photocopying and other methods. Enough information is left out to allow the users to meet their specific needs, adapt to their climate and readily available building materials and methods. The design configuration basically works as an inclined tube, open at both ends. The heat-generating spaces are grouped at the upper open end and the comfort spaces at the lower open end. This allows redundant effects to work together to achieve good air-change in the interior, and the segregation of good and bad air. With its basic configuration being the key to sustainability, widespread applications of Bayanihan can reduce energy consumption, enhancing people's awareness of the advantages of using sustainable local building methods and thus having a beneficial long-term impact on the environment.

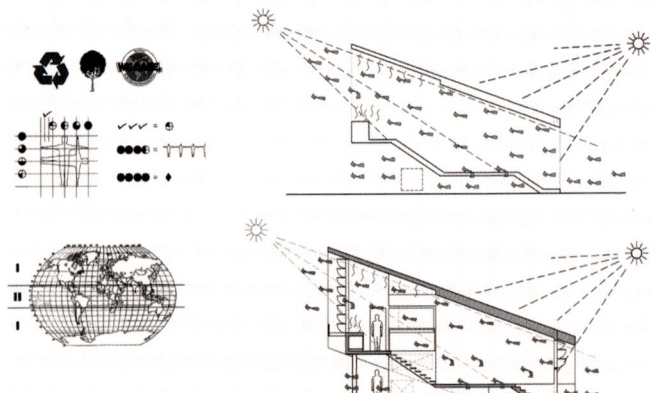

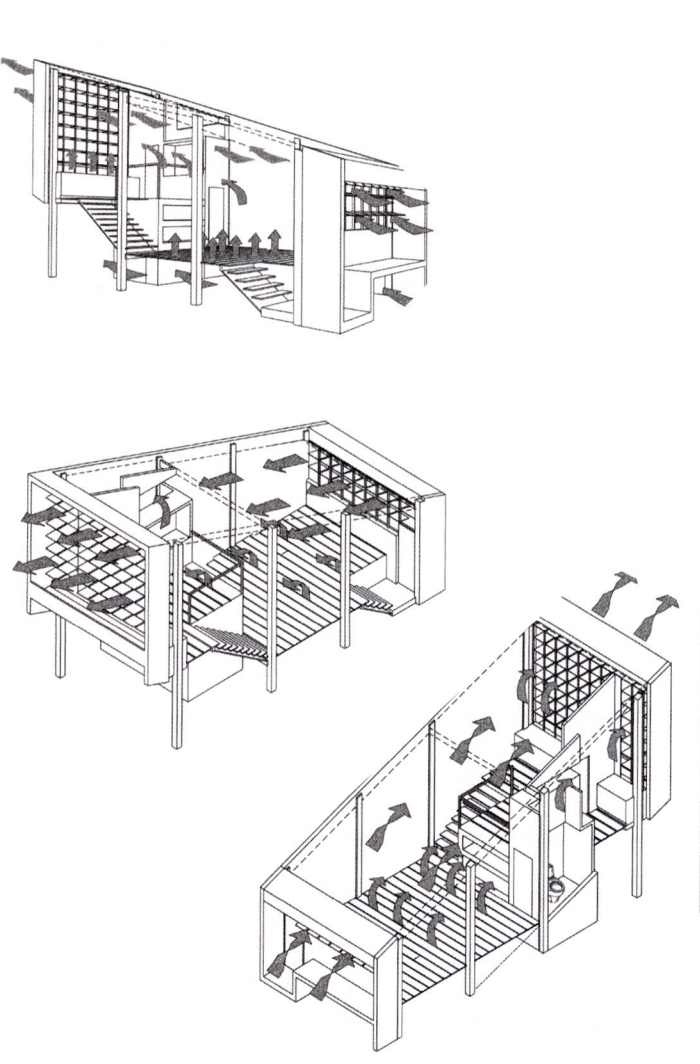

DO IT YOURSELF

AUTO- & BIOGENERATED

MATERIALS & CONSTITUTION

RECYCLED & READAPTED

MODULATED & PRE-ARRANGED

DIGITAL FAB

IEVGENII, Vasiliev   PGASA - Academy of Architecture

archi427@gmail.com

In the same way that a child plays with a Rubik's Cube, with all of a child's delight, absorption and interest, so the occupants create their own house. The space for creative ideas is thus practically unlimited within the bounds of this project. It is also as simple as one, two, three, four, five. Each element of a Rubik's Cube is a module, and the occupant-creators decide whether these correspond to their way of life. A number of functional modules are available: the sleeping module, the vestibule-kitchen module, the children's module, the entertaining module and the work module. The sleeping and vestibule-kitchen modules are two basic necessities for domestic life, and for these a variation in layout is offered, with this layout also being created by the occupants of the house. All of the essential elements for domestic life can be easily assembled and dismantled, so the dwelling can change its orientation or be moved from place to place. Assembly of the module will take the occupants roughly a quarter of an hour, with not much physical effort, on their own or with the help of others. The Rubik's Cube can be integrated into either a natural environment or a city without difficulty, thanks to the fact that the façade is designed to be modulated. There is no limit to the number of modules that can be incorporated into a cube: the only limits are those imposed by the conditions of the site where the project is implemented. The Rubik's Cube is completely independent, but can also be connected to the public utilities.

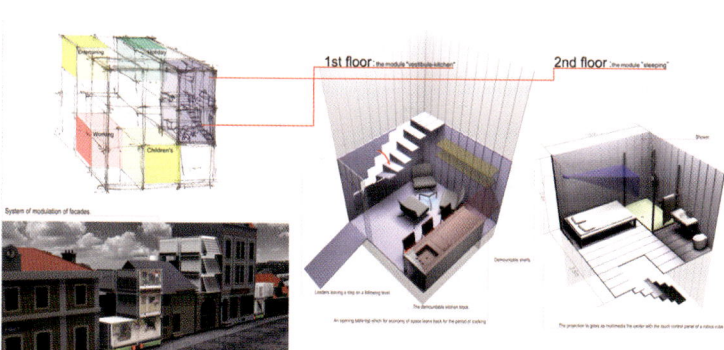

DO IT YOURSELF

AUTO- & BIOGENERATED

MATERIALS & CONSTITUTION

RECYCLED & READAPTED

MODULATED & PRE-ARRANGED

DIGITAL FAB

System of modulation of facades.

1st floor : the module "vestibule-kitchen"

2nd floor : the module "sleeping"

Emergency

Washing

Working

Children's

Shower

Ladders serving a step on a following level

Demountable shaft

The demountable kitchen block

An opening tabletop which for economy of space leans back for the period of cooking

The projector is gives as multimedia the center with the touch control panel of a colour cube

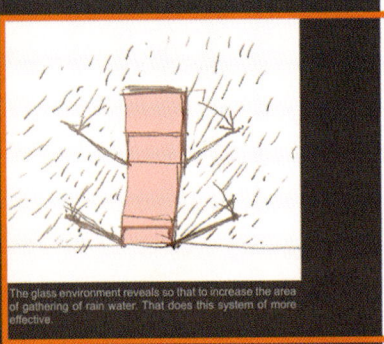

The glass environment reveals so that to increase the area of gathering of rain water. That does this system of more effective.

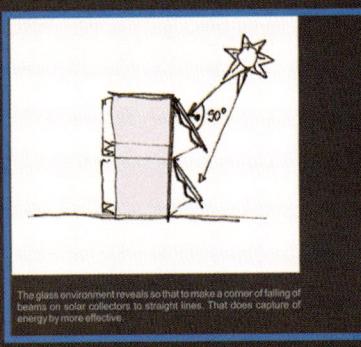

The glass environment reveals so that to make a corner of falling of beams on solar collectors to straight lines. That does capture of energy by more effective.

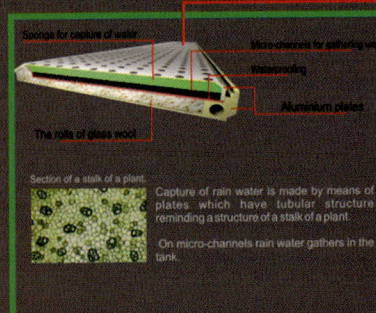

Sponge for capture of water

Micro-channels for gathering water

Waterproofing

Aluminium plates

The rolls of glass wool

Section of a stalk of a plant.

Capture of rain water is made by means of plates which have tubular structure reminding a structure of a stalk of a plant.

On micro-channels rain water gathers in the tank.

Id:511b13

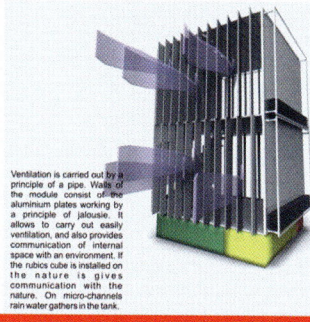

Ventilation is carried out by a principle of a pipe. Walls of the module consist of the aluminium plates working by a principle of jalousie. It allows to carry out easily ventilation, and also provides communication of internal space with an environment. If the rubics cube is installed on the nature is gives communication with the nature. On micro-channels rain water gathers in the tank.

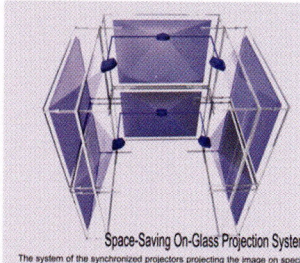

Space-Saving On-Glass Projection System

The system of the synchronized projectors projecting the image on special glass enables to be arranged under an environment, under existing city building not breaking its integrity, and also enables the tenant most to model external shape of the house. This system can be used also as multimedia center for family viewing video. Probably also its use in commercial objectives, as advertising panels.

The glass environment the Glass environment creates a special microhabitat, thus not breaking communication of internal space of the house with the nature.

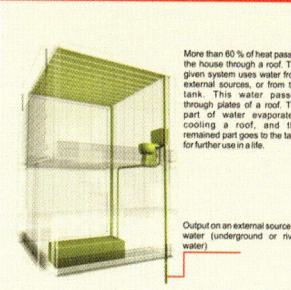

More than 60 % of heat pass in the house through a roof. The given system uses water from external sources, or from the tank. This water passes through plates of a roof. The part of water evaporates, cooling a roof, and the remained part goes to the tank for further use in a life.

Output on an external source of water (underground or river water)

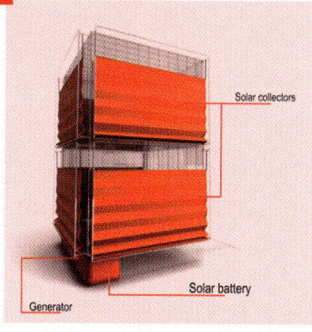

Solar collectors

Solar battery

Generator

DO IT YOURSELF

AUTO- & BIOGENERATED

MATERIALS & CONSTITUTION

RECYCLED & READAPTED

MODULATED & PRE-ARRANGED

DIGITAL FAB

SAN GREGORIO, Sara

Escuela Técnica Superior de Arquitectura
de Madrid (ETSAM)

sarasg@interlink.es

The project proposes a new approach to the construction of a house, based on the setting up of a company that will supply the construction materials, the systems, the elements and the tools, together with the instructions for the construction to be carried out.

The project proposes self-constructible habitable segments. The system of self-construction consists of industrialized components with which self-supporting units can be created, based on a structure of articulated rods that support the façade materials together with the floor, ceiling and partition walls.

A system of progressive growth is proposed, allowing the creation of simple

units that grow by grouping. This type of space, created by aggregation, allows different types of accommodation and different ways of living and living together. A set of grouped modules can always be returned to a deblock or a new module can be added when necessary.

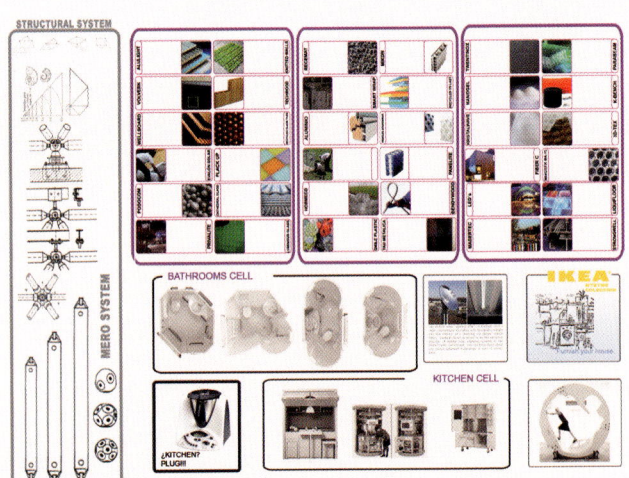

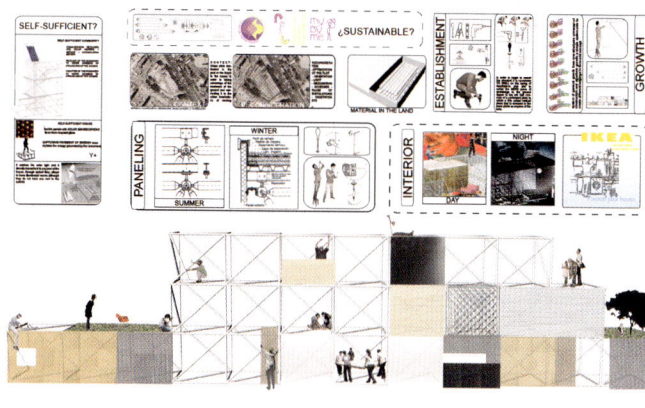

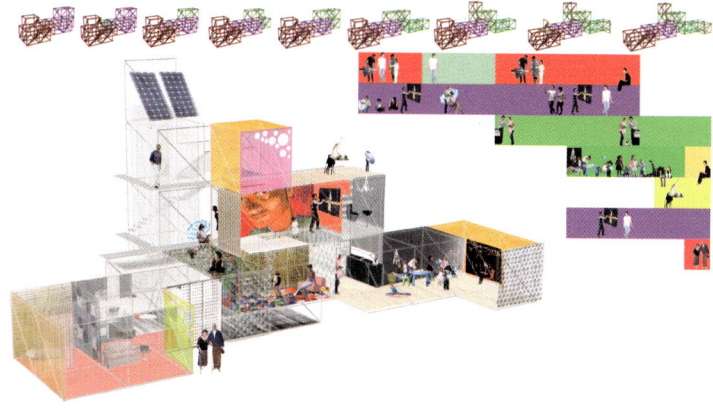

DO IT YOURSELF

AUTO - & BIOGENERATED

MATERIALS & CONSTITUTION

RECYCLED & READAPTED

MODULATED & PRE-ARRANGED

DIGITAL FAB

SIERRA DE HITA, Luis

Escuela Técnica Superior de Arquitectura de Madrid (ETSAM)

indioapatxi@hotmail.com

1. Location

The building is located in a supermarket car park in Alcobendas, Madrid.

2. Sustainability and self-fab

The project aims to occupy a place that is already occupied and obtain the maximum benefit from the whole area. The building should work throughout the day and the activities themselves should be carried out without any specific order, with the functional layers being superposed. The building is created as a machine, a tool that only works when the user is there. This idea allows the self-fab use to be established and finally implemented in the building. Each tool, called a sprout, depends not only on the user but also on the other sprouts in order to function properly.

3. Building Structure

The assembly of the structure is represented in inverse order because recycling is an essential aspect of the project.

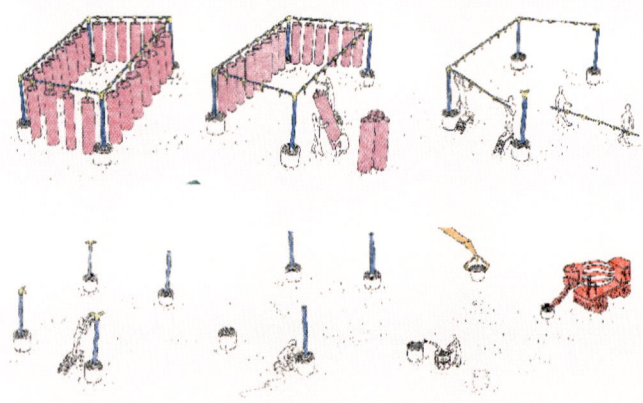

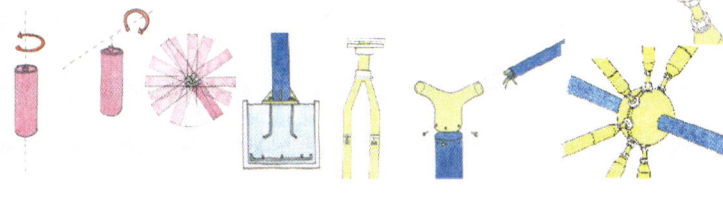

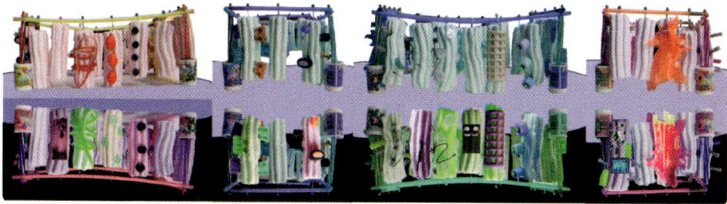

DO IT YOURSELF

AUTO- & BIOGENERATED

MATERIALS & CONSTITUTION

RECYCLED & READAPTED

MODULATED & PRE-ARRANGED

DIGITAL FAB

b.4 PREMOLYAL HIDING PLACE

1) USAGE
This sprout allows avoid inquisitive looks for a short period of time.

1.2 WORKING
A base structure with velcro in its interior perimeter forms this sprout. The user must complete this structure adding the hiding material. Lightness and velcro union are the only requirements. The lowliness of the material is valued positively.

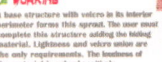

b.12 RADIANT MATERIAL

b.5 RENTABLE CLOTHING HORSE

1) USAGE
This sprout is usually used to hang out the clothes after taking a shower. Natural sounds are produced when drying the clothes are hung.

1.2 WORKING
A folding structure forms this sprout. Inside the sprout there are clothes pegs. In order to achieve a more effective drying, the sprout can be elevated.

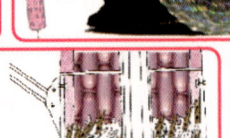

b.6 SCROLLABLE MATERIAL AND INSULATOR

1) USAGE
This sprout allows cover the whole building structure in order to create intimate spaces. Intimacy range depends on the material properties.

1.2 WORKING
Inside the... different material...

b.7 CUSTOMIZED GRASSBED

1) ESTRUCTURAL BASE
It is necessary a crate with gas on the bottom.

1.2 SUBSTRATUM
The crate must be covered with a waterproof sheet and fill in with substratum.

1.9 MOULD CREATION
Above the substratum, the user must place the seeds. Over them a phonetic sheet.

1.4 PERSONALIZE
The user must lie down and because of the phonetic sheet memory, its body form stays.

b.8 NATURAL TOILET

1) USAGE
This sprout functions as a temporary toilet.

1.2 WORKING
Toilets are indicated by folding circular elements. These elements are unfolded when the user is in their interior.

grass
sand
gravel
geotextile sheet
drainage tube
waterproof sheet
occupied toilet
not occupied toilet

b.9 AROMATIC FANS

1) USAGE
This sprout not only improves the environmental conditions but also allows personalize its smell for a certain time.

1.2 WORKING
It is necessary press a button to activate aromatic fans. Scents can be manually placed by the user. The creation of a perfume workshop is considered.

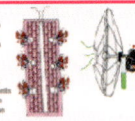

b.20 ENERGETIC WINDMILLS

b.10 MOVING CINEMA

1) USAGE
This sprout can be used only during the evening to show videos or slides.

1.2 WORKING
A selfscrollable screen is located on the top of the estructure. Rentable machines are placed on the bottom of the estructure.

b.21 RENTABLE CAMPING POINTS

b.11 PHOTOVOLTAGE PANELS

1) USAGE
Sun energy is captured during the daytime and accumulated in order to be used later.

1.2 WORKING
Machine working of other sprouts depend on this energy.

b.22 HANGING BED
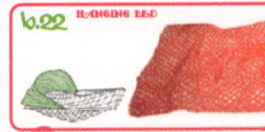

DO IT YOURSELF

AUTO- & BIOGENERATED

MATERIALS & CONSTITUTION

RECYCLED & READAPTED

MODULATED & PRE-ARRANGED

DIGITAL FAB

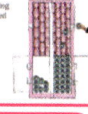

...FACE

U USAGE
...sprout functions as a CD recycling ... The compact discs are used ... to create protected shells.

WORKING
Compact discs are joined by ... cards. The creation of a ... workshop is considered.

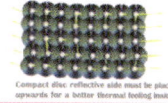

Compact disc reflective side must be placed upwards for a better thermal feeling inside.

6.15 INTERACTIVE RECORDER AND STORY TELLING

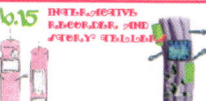

U USAGE
This sprout functions as a recorder of sound, especially stories and messages.

WORKING
Recorder machines can be used free of charge.

...TER BEDS

U USAGE
Water beds can be hired for some hours. The user chooses the temperature of the water.

WORKING
This sprout depends on the building because of the necessity of water.

6.3 TEMPORARY SHOWERS AND SPRINKLERS

U USAGE
The sprinklers improve the environmental conditions and water the grass beds. The showers can be used not only by person but also by cars.

WORKING
The working of the sprout depends on the water reserves.

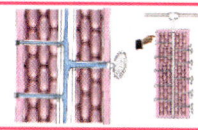

6.19 CHILDREN GAMES AND FLUORESCENCES

U USAGE
While parents shop in the supermarket, children use this sprout as a game.

WORKING
The sprout is totally covered by velcro. Children create different fluorescent figures during the day. The fluorescences light the evening activities.

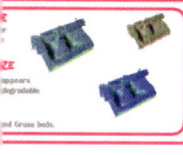

...ZE

...appears

...degradable.

...and Grass beds.

6.14 RENEWABLE ELECTRICAL SOCKETS

U USAGE
This sprout allows the electrical preparation of the different activities.

WORKING
Sockets can be extended if the user pull them.

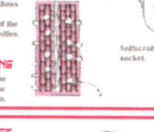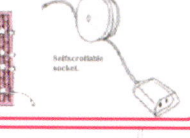

Self-scrollable socket.

6.16 FOOD HEATER

U USAGE
This sprout functions as a food heater.

WORKING
Food is heated up in a container which is covered with hot water.

food

hot water

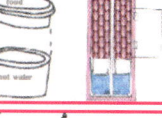

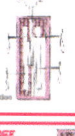

6.1 RAIN RECEIVER AND PARASOL

U USAGE
This sprout not only captures water but also provides temporary protection against rain or sun.

WORKING
Water is collected by an umbrella structure and it is carries through the building structure to other sprouts.

...USAGE

...sprout allows ... use of hot water.

...WORKING

...size tapping ...form the sprout, ...cars or ...grass beds ...se of the ...e sons.

6.2 HYDROPONIC CROPS AND RENEWABLE LAMPS

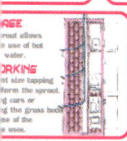

U USAGE
One of the side of the sprout is used to grow hydroponic crops and the other side is used as a temporary lighting.

WORKING
The use of the routable lamps is free but the user must add the light bulb.

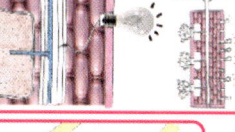

...is located on the top of ... and it is used as a ....

...NG

...telescopic structures ...ry for use this sprout. ...sewn on these structures.

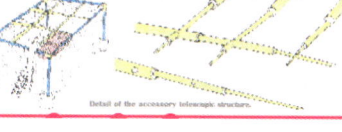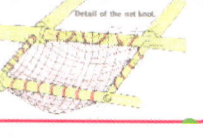

Detail of the net knot.

Detail of the accessory telescopic structure.

69B97D

RAMMED EARTH TOWER

Japan

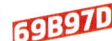

YOSHIZAWA, Ikuma
MIYAWAKI, Ryohei
YU, Lin Shuan yu,

AMORPHE Takeyama & Associates

ikuma125@process5.com

Until now we have developed our civilization. Global warming. Industrialization. Commercialization. Depletion of the ozone layer. The problem of waste. Water pollution. The rise in the level of the sea. Poverty. Population increase for whom, and for what? Mature technologies and a culture that were developed throughout the history of civilization: Why not use these effectively in nature, just for ourselves? Why not live our lives in harmony with this great planet, just for ourselves? We would like to propose a kind of Living Unit, it creates the community that centres on soil. We can construct our homes with our families, a little at a time, day by day. It will not be difficult, but it will need some time to grow. One day it will become a new village, a new town, and then it will be a new culture.

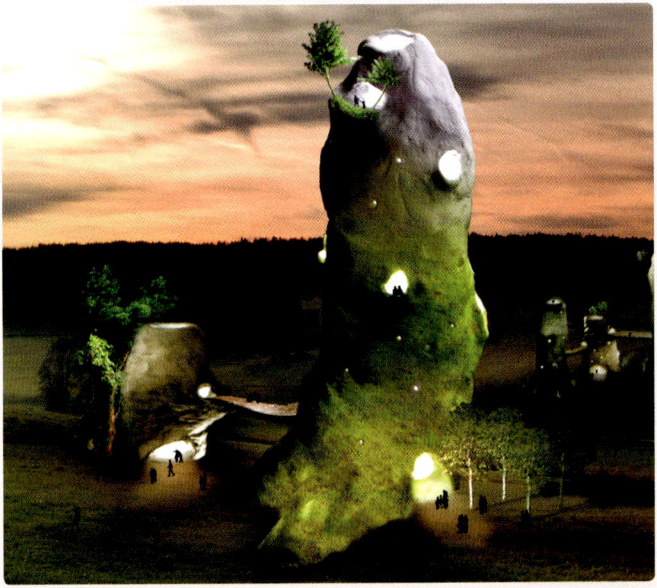

DO IT YOURSELF

AUTO- & BIOGENERATED

MATERIALS & CONSTITUTION

RECYCLED & READAPTED

MODULATED & PRE-ARRANGED

DIGITAL FAB

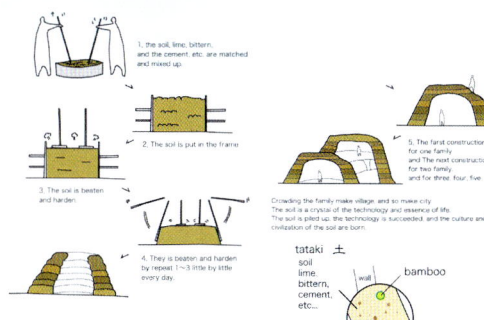

1. the soil, lime, bittern, and the cement, etc. are matched and mixed up

2. The soil is put in the frame

3. The soil is beaten and harden

4. They is beaten and harden by repeat 1~3 little by little every day.

5. The first construction is for one family and The next construction is for two family and for three, four, five.

Crowding the family make village and so make city
The soil is a crystal of the technology and essence of life.
The soil is piled up, the technology is succeeded, and the culture and the civilization of the soil are born.

tataki 土
soil
lime
bittern,
cement,
etc...

bamboo

wall

wall

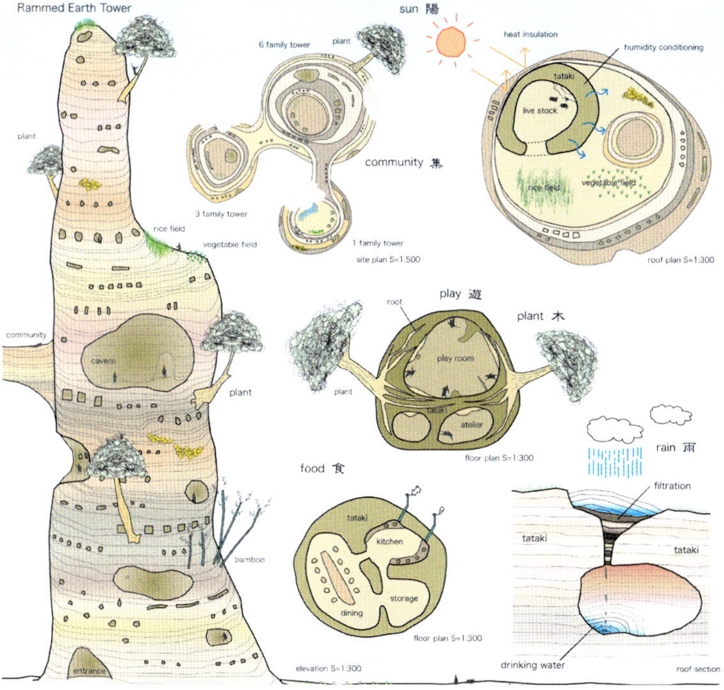

Rammed Earth Tower

plant

plant

rice field

community

cavern

plant

bamboo

entrance

elevation S=1:300

sun 陽

6 family tower   plant

community 集

3 family tower

vegetable field

1 family tower

site plan S=1:500

heat insulation

humidity conditioning

tataki

live stock

rice field

vegetable field

roof plan S=1:300

play 遊

root

play room

plant

plant

atelier

floor plan S=1:300

plant 木

food 食

tataki

kitchen

storage

dining

floor plan S=1:300

rain 雨

filtration

tataki

tataki

drinking water

roof section

59

·Soil and the sun 陽

The sun is living thing's energy.
The plant is grown, and oxygen is invented.
However, the sun warms the earth.
The soil defends people who live internally from heat.
Moreover, the soil defends the heat that tries to be deprived the outside when it is cold.

·Soil and food 食

Food is hope to the life.

animal : the animal.
The feeding chain happens on the earth.

·Play with the soil 遊

It enjoys running about the earth, and competing.
The person gets the energy of the name of playing from the earth.
The soil is learnt, and it digs, it mixes up, it piles up, and a lot of beaten play is learnt from the soil.
From the adult to the child , play is a means to tell the technology.

集

DO IT YOURSELF

AUTO- & BIOGENERATED

MATERIALS & CONSTITUTION

RECYCLED & READAPTED

MODULATED & PRE-ARRANGED

DIGITAL FAB

BARCO, María Anton

Universidad CEU San Pablo

m_anton__arq@hotmail.com

Our project was born with an important premise: the need for a bioclimatic building. As time goes by, vegetation starts to dominate the building and its structure, which eventually disappears.

The material used is a synthesis of waste plastics, from which each of the elements needed for the house is configured. We propose a Self-Fab house made of biodegradable plastic, which can simply disappear when it is no longer required. As a result we are going to RE_WRITE this exercise, designing not only a self-sufficient house but one that must be RE_CYCLED. Although our society is becoming more and more conscious of the real need to stop wasting energy, nobody seems to see that the biggest problem is not the excessive consumption of energy but the generation of waste.

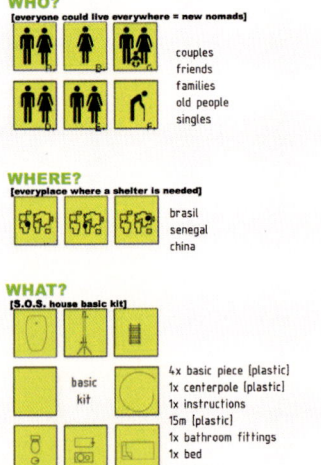

WHO?
[everyone could live everywhere = new nomads]

couples
friends
families
old people
singles

WHERE?
[everyplace where a shelter is needed]

brasil
senegal
china

WHAT?
[S.O.S. house basic kit]

basic kit

4x basic piece [plastic]
1x centerpole [plastic]
1x instructions
15m [plastic]
1x bathroom fittings
1x bed

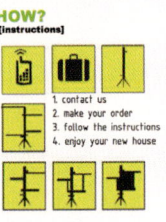

HOW?
[instructions]

1. contact us
2. make your order
3. follow the instructions
4. enjoy your new house

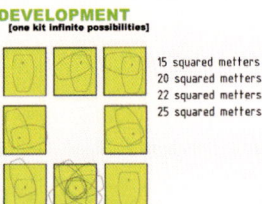

DEVELOPMENT
[one kit infinite possibilities]

15 squared metters
20 squared metters
22 squared metters
25 squared metters

DO IT YOURSELF

AUTO- & BIOGENERATED

MATERIALS & CONSTITUTION

RECYCLED & READAPTED

MODULATED & PRE-ARRANGED

DIGITAL FAB

BRASIL

SURFACE: 8.514.877 km2
H2O %: 0,65%

POPULATION [5th PLACE]
· TOTAL
188.098.127 (1/7/2006
· DensiTY
22 hab/km2

CHINA

SURFACE: 9.596.960[2] km2
H2O %: 2,82%

POPULATION [1º PLACE]
· TOTAL
1.313.973.713 (2006 est.)
· DensiTY
136,12[2] hab/km2

SENEGAL

SURFACE: 196.190 km2
H2O %: 2,1%

POPULATION [70th PLACE]
· TOTAL
11.987.121 (2006 est.)
· DensiTY
52 hab/km2

DO IT YOURSELF

AUTO- & BIOGENERATED

MATERIALS & CONSTITUTION

RECYCLED & READAPTED

MODULATED & PRE-ARRANGED

DIGITAL FAB

BERGERON, Gabriel
WEST, Jennifer
BOEHM, William

Boehm Architecture

gabe@boehmarchitectrue.com

We live in the Consumer Age. Human life has been relegated to a combination of isolation within the home and consumption-based social exposure. Entertainment replaces interaction. The virtual replaces the physical. Shopping replaces making, thinking and learning.

Our focus is the BIGBOX store. Typical business models use a 10-year life cycle, leaving a series of abandoned BIGBOX shells and parking lots. We envision a new development model that transforms vacant BIGBOX shells and their surrounding parking lots into centres for user-created and maintained communal settlements. The BIGBOX carcass becomes a building resource centre, containing workshops, material supply areas, expert housing, classrooms, and community gathering spaces. Housing units are assembled through the cooperation of a group of neighbours. Visiting experts provide guidance to help the local people build one another's houses.

The housing units will be organized around small-scale community courtyards that will act as utility connections and vertical circulation cores. The units will be composed of a combination of pre-manufactured panel units that allow for ad-hoc additions using any material or detail that the users would like to add to their homes. The resulting form will be a combination of high-technology cores and idiosyncratic additions and personal expressions.

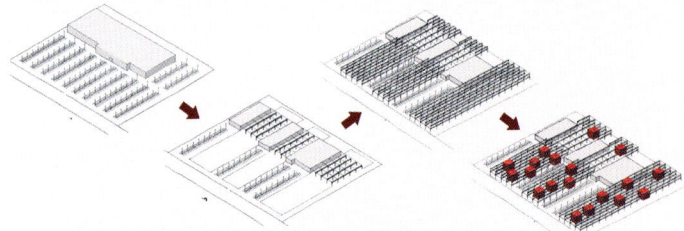

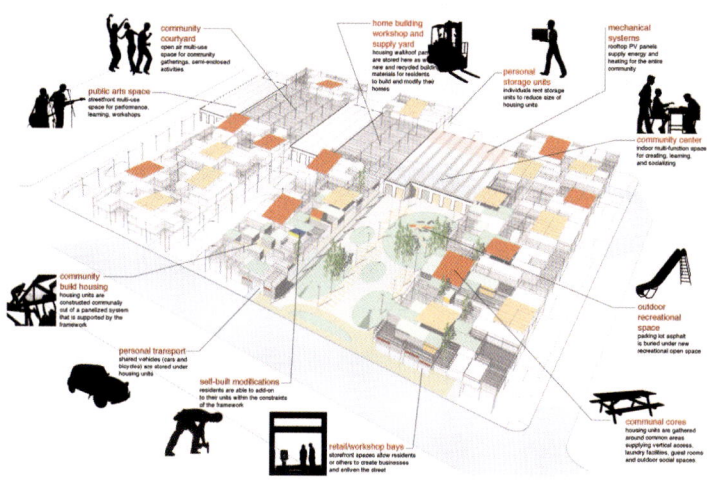

community courtyard
open-air multi-use space for community gatherings, semi-enclosed activities

public arts space
storefront multi-use space for performance, learning, workshops

home building workshop and supply yard
housing walkout panels are stored here as a new and recycled building materials to build and modify their homes

personal storage units
individuals rent storage units to reduce size of housing units

mechanical systems
rooftop PV panels supply energy and heating for the entire community

community center
indoor multi-function space for creating, learning, and socializing

community built housing
housing units are constructed communally out of a panelized system that is supported by the framework

personal transport
shared vehicles (cars and bicycles) are stored under housing units

self-built modifications
residents are able to add-on to their units within the constraints of the framework

retail/workshop bays
storefront spaces allow residents or others to create businesses and enliven the street

outdoor recreational space
parking lot asphalt is buried under new recreational open space

communal cores
housing units are gathered around common areas supplying vertical access, laundry facilities, guest rooms and outdoor social spaces.

after

DO IT YOURSELF

AUTO- & BIOGENERATED

MATERIALS & CONSTITUTION

RECYCLED & READAPTED

MODULATED & PRE-ARRANGED

DIGITAL FAB

C8F35C

PRESSGROVE, David

ARC + CON

nltlu@yahoo.com

In August 2005, thousands of homes on the Biloxi peninsula were severely damaged or completely destroyed as a result of Hurricane Katrina. The entire city was piled high with building debris. Barges, boats and ships were swept ashore, wiping out huge swathes of ancient oak trees that had long served as hurricane buffers. No one knew if the homes should be rebuilt, or, if so, how.

At first it seems a bad idea to build in an area so susceptible to hurricanes. As a wetland, the area should be conserved in its pristine state, as barrier islands are. But on second thoughts, having got to know a place, a community with a history and a culture, perhaps we do have to rebuild here, with an architecture strong enough to withstand gale-force winds and durable enough to survive submersion. The third consideration is that there is very little money. The Federal Government has issued Revised Flood Plain Levels, raising the height at which many families will have to build by 3, 6, and 8 metres. The insurance companies cancelled homeowners' policies, exploiting technicalities to renege on billions of dollars worth of flood insurance claims, and some pulled out of the area altogether. The Government began to compensate homeowners for the increased costs of higher, stronger housing. Many did not qualify. Non-governmental organizations began issuing grants as well — but, again, many didn't qualify. Thousands of families are still waiting, two years after Hurricane Katrina. The City of Biloxi spent more than two years and $ 81.5 million removing 2.98 million cubic yards of debris from the city — enough debris to cover a football field and stand 139.7 stories high.

It was said that reusing existing building materials was unsafe, that the mould and mildew would be unhealthy. But every piece of lumber dried out, and all the mould and mildew died with ventilation, dry conditions, and exposure to sunlight. Knowing this, many families collected as much of these materials as they could for reuse.

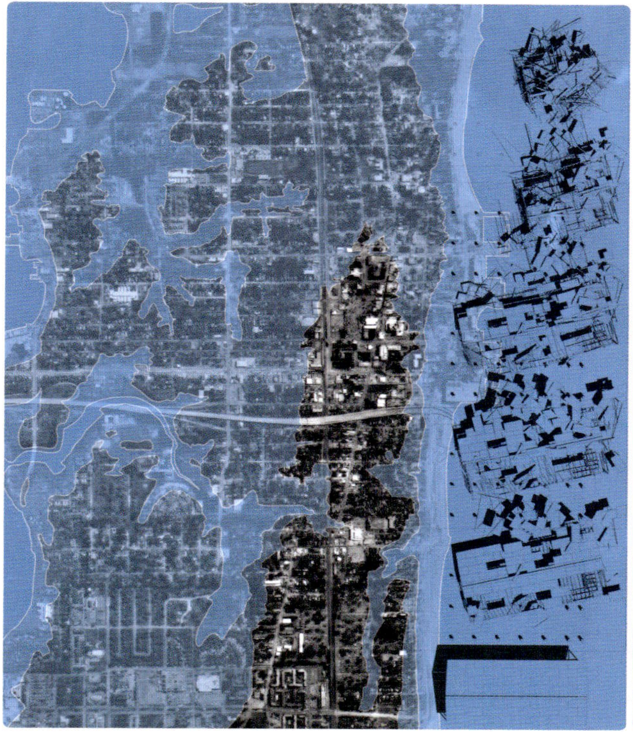

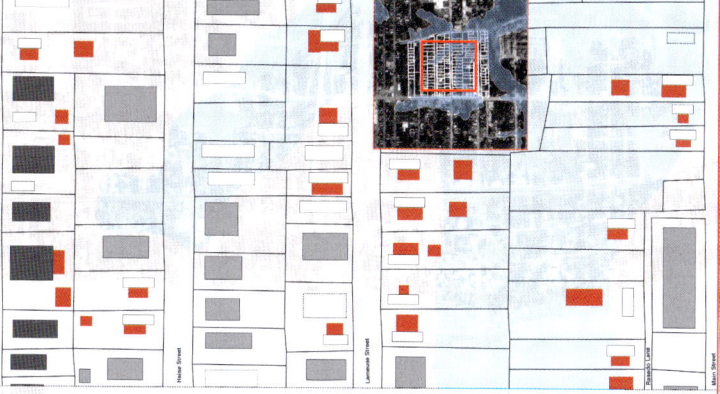

DO IT YOURSELF

AUTO- & BIOGENERATED

MATERIALS & CONSTITUTION

RECYCLED & READAPTED

MODULATED & PRE-ARRANGED

DIGITAL FAB

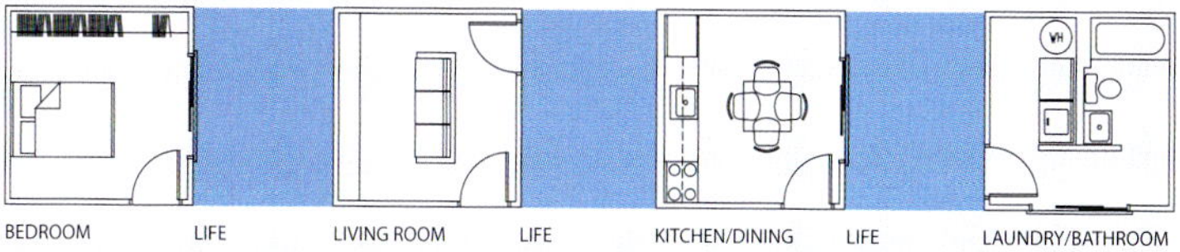

DO IT YOURSELF

AUTO- & BIOGENERATED

MATERIALS & CONSTITUTION

RECYCLED & READAPTED

MODULATED & PRE-ARRANGED

DIGITAL FAB

BEDROOM   LIFE   LIVING ROOM   LIFE   KITCHEN/DINING   LIFE   LAUNDRY/BATHROOM

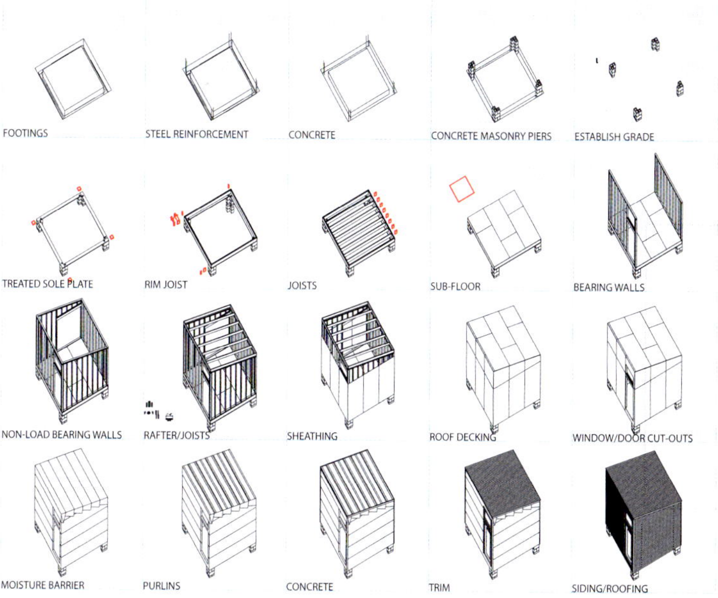

FOOTINGS   STEEL REINFORCEMENT   CONCRETE   CONCRETE MASONRY PIERS   ESTABLISH GRADE

TREATED SOLE PLATE   RIM JOIST   JOISTS   SUB-FLOOR   BEARING WALLS

NON-LOAD BEARING WALLS   RAFTER/JOISTS   SHEATHING   ROOF DECKING   WINDOW/DOOR CUT-OUTS

MOISTURE BARRIER   PURLINS   CONCRETE   TRIM   SIDING/ROOFING

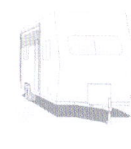
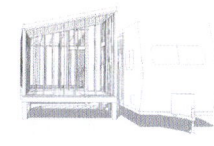
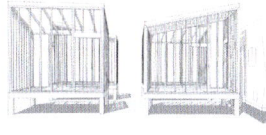

MEZOUED MOUAD, Aniss
SEBKI, Toufik

EPAU

aniss.mezoued@gmail.com

One of the crucial problems that the future holds for humanity, apart from population growth, energy consumption and dwindling resources, is major climate change, and particularly global warming. With the desertification of the planet as one of the main consequences, we take a keen interest in arid areas, and specifically in one of the largest deserts on the planet, the Sahara. Human settlement there has always been in harmony with the environment and climate, but with an imbalance between population growth and the exploitation of resources, leading the authorities to ban the use of palm trees in construction. It being no longer possible to cut down palm trees in places such as Taghit, in the south-west of Algeria, new construction techniques, inappropriate to the environment, were adopted. Earth and wood have been replaced by concrete and brick, resulting in buildings hopelessly unsuited to the desert conditions and consuming huge quantities of energy for heating and air-conditioning.

Steps of the new system

Installation of beams

Construction of earth walls between the beams. That permits the horizontally wind bracing between porticos.

Finishing of walls by spattered mortar made with stabilized earth.

DETAILS

Fabrik

Solar panel

Air-conditioning is natural, with the earth mass and the system of air circulation and heat evacuation.
The heating, used only in winter nights is realized thanks to solar energy or like in the old days, by burning dried cow pat.

Water is stocked in tanks arranged between porticos.

For freshness of water, we cover the tank by earth, which will reduce its exposure to external heat.

For hot water, the tank is exposed to sun rays in the morning, and reheated during winter by solar panel arranged on the roof.

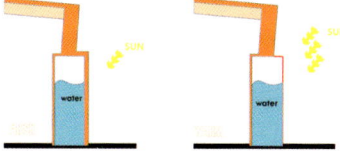

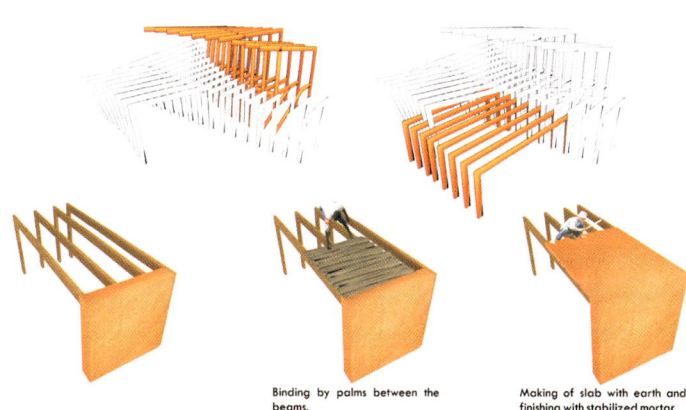

Binding by palms between the beams.

Making of slab with earth and finishing with stabilized mortar.

DO IT YOURSELF

AUTO- & BIOGENERATED

MATERIALS & CONSTITUTION

RECYCLED & READAPTED

MODULATED & PRE-ARRANGED

DIGITAL FAB

YENERICH, Adan
SCHIAVONI, Rodrigo

Yenerich Adan, Arquitecto
Schiavoni Rodrigo, Arquitecto

adanyenerich@hotmail.com

Primary Environmental Reality

Santiago del Estero once had Xxxxxx km² of native forest, with a small population living in a balanced relationship of supply and maintenance between man and forest.

Present Environmental Reality

From the 1980s on the expansion of agricultural land has greatly increased the clearing of trees and the loss of 39,180 km² of native forest. The biodiversity of the ecosystem was destroyed and the local population migrated to the big cities. The process of desertification caused the salinization of the fragile soil of the primitive forest, as a result of which 53% of the local farmers had to abandon cultivation. The outcome of this devastating process is that 10% of the land is in a state of severe desertification, with a further 60% in a state of moderate desertification.

Regenerated Environmental Reality

Our project aims to create a process of regeneration of the forest through the relocation of settlers. The instrument to obtain it we denominated it:

BUFR > Basic Unit of Forest Recovery = REGENERATION OF NATIVE FOREST

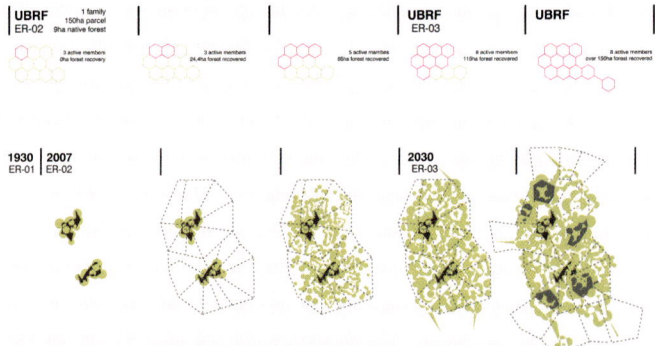

DO IT YOURSELF

AUTO- & BIOGENERATED

MATERIALS & CONSTITUTION

RECYCLED & READAPTED

MODULATED & PRE-ARRANGED

DIGITAL FAB

Climatic bejavior

Summer med27°C / max47°C - rain 750mm/y   Winter med12°C /min-5°C

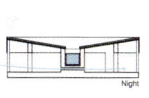 Night

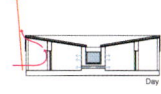 Day

 Rain

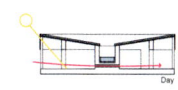 Day

Components

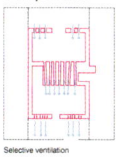
Selective ventilation

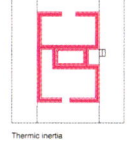
Thermic inertia

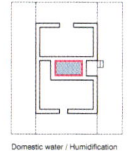
Domestic water / Humidification

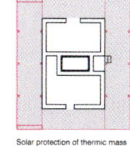
Solar protection of thermic mass

Retraining and energy

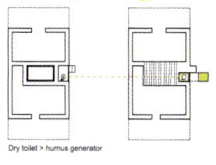
Dry toilet > humus generator

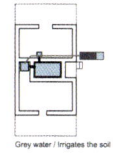
Grey water / Irrigates the soil

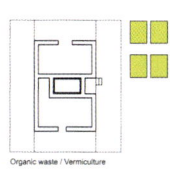
Organic waste / Vermiculture

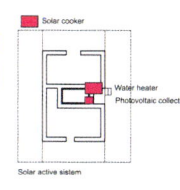
Solar cooker
Water heater
Photovoltaic collector
Solar active sistem

Materials

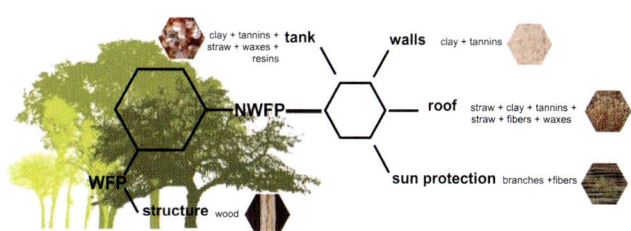

clay + tannins + straw + waxes + resins — **tank**

walls clay + tannins

NWFP

roof straw + clay + tannins + straw + fibers + waxes

WFP

sun protection branches +fibers

structure wood

FOREST REGENERATOR

F35501
SELF-SUFFICIENT SINGLE HOUSING

UBRF
ER-02

UBRF
ER-03

DO IT YOURSELF

AUTO- & BIOGENERATED

MATERIALS & CONSTITUTION

RECYCLED & READAPTED

MODULATED & PRE-ARRANGED

DIGITAL FAB

NEWAZ KHAN, Nabi
MAJID ASADUZZAMAN, Lutfullahil

Archeground

shomin2698@yahoo.com

We see many houses in nature. Our intention was to take the essence of that very houses which prevail in nature and are undoubtedly sustainable. We have selected bee-comb, tailor bird's nest and termite mound for their significant features. From bee-comb, we have tried to interpret its structural system and spatial space organization. Termite mound has attracted us for its shape [form] and the basic material mud. The tailor bird's nest has been considered important for its texture or fabrication. Our house is made of mainly mud mixed with hay for more strength , bamboo for structure, jute fabric, bamboo silka for roof fabrication.

DO IT YOURSELF

AUTO- & BIOGENERATED

MATERIALS & CONSTITUTION

RECYCLED & READAPTED

MODULATED & PRE-ARRANGED

DIGITAL FAB

Series of hut

seed's bed

Series of termite mound

b a n g l a d e s h

green

termite mound >

u s e   i n   n a t u r e >>

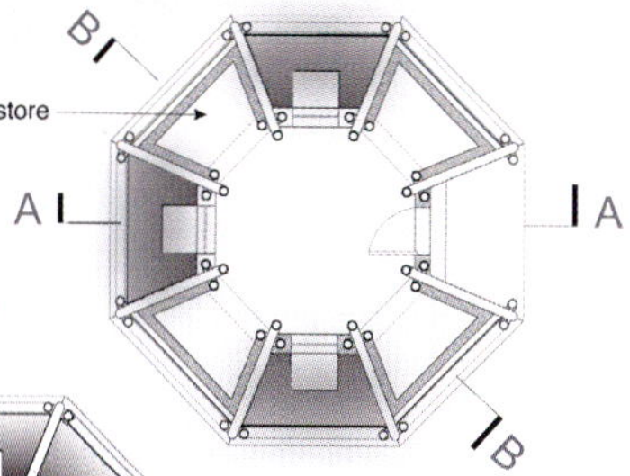

store

B|

A| |A

|B

plan at + 2.5'
[Not in scale]

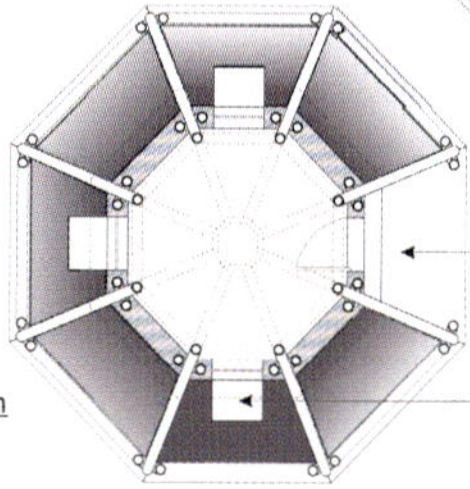

Entry

window

plan at + 5'
[Not in scale]

c o n s t r u c t i o n   p r o c e s s >>

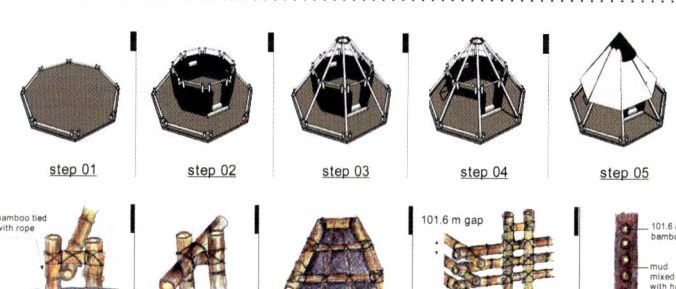

step 01

step 02

step 03

step 04

step 05

bamboo tied with rope

detail at **A**

detail at **B**

detail at **C**

101.6 m gap

detail at **D**

101.6 mm bamboo

mud mixed with hay

detail at **E**

KATSIKIS, Nicolaos
IASON, Pantazis

NTUA

wizy82@gmail.com

How we do it:

The cardbin consists of a series of selfcarrying structural components which serve also as its skin.

These parts are pre-designed so as to propose a simple construction system but nothing more than that: How you assemble them is totally defined. How you compose them is definitely not. However a generic type of house is presented so as to demonstrate the efficiency of the system.

But as a matter of fact the cardbin is not a house. It is a process.

During this process folding plays a basic role: The cardbin's components as the cardbin itself are also self fabricated. Due to the reason that rapid and flexible construction requires rapid and flexible transportation of the structural materials, the cardbin uses folded heavy-duty cardboard enclosures which can be cast on site using locally produced cement, plaster, resins or even soil [allowing a green roof or skin to grow on it].

Except from its light weight, heavy-duty white cardboard is also foldable and strong enough to be used as this peculiar kind of origami envelope. What's more it guarantees low cost and it is 100% recyclable. By folding, casting and assembling the components one after an other the user can feel free to experiment with the synthesis of form while the construction method enables endless assembling and disassembling.

DO IT YOURSELF

AUTO- & BIOGENERATED

MATERIALS & CONSTITUTION

RECYCLED & READAPTED

MODULATED & PRE-ARRANGED

DIGITAL FAB

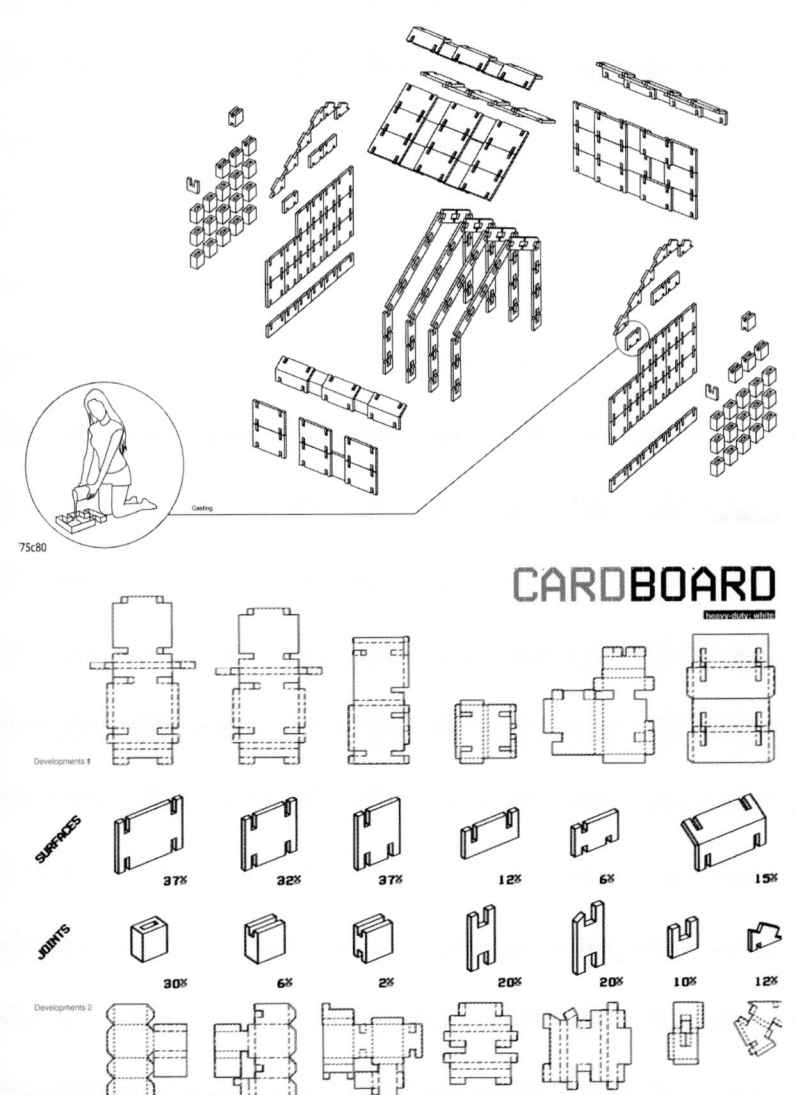

75c80

Casting

CARD**BOARD**

heavy-duty, white

Developments 1

SURFACES

37% 32% 37% 12% 6% 15%

JOINTS

30% 6% 2% 20% 20% 10% 12%

Developments 2

DO IT YOURSELF

AUTO- & BIOGENERATED

MATERIALS & CONSTITUTION

RECYCLED & READAPTED

MODULATED & PRE-ARRANGED

DIGITAL FAB

AUTOGENERATED & BIOGENERATED

Considering the fact that the buildings as long as the cities themselves are living and evolutionary organisms, Auto-Generated and Bio-Generated, is exploring the possibilities offered for the development of a more sustainable and self sufficient-housing.

The new computational and scientific advances could be used as tools in order to simulate the ways the nature is developing, evolving and functioning. Moreover, the ways the people (inhabitants) are behaving could be predicted, so as for the options of city development to be generated and the developing rules to be defined.

Additionally, the buildings themselves could imitate natural functions as long as the way living organisms are developing and grow. This could result in the better adaptation of the building in different environments and its development as a more autonomous entity.

Once we are talking about Auto-Generation and Bio-Generation we are referring both to the possibility of using the mechanical advances in order for the building to be auto-

DO IT YOURSELF

AUTO- & BIOGENERATED

MATERIALS & CONSTITUTION

RECYCLED & READAPTED

MODULATED & PRE-ARRANGED

DIGITAL FAB

constructed, as long as the use of certain Bio-Chemical processes where the building could grow itself by self-reproducing its own molecules. In this case certain parts of the building could collectively create and produce new ones; the structure is being developed by its own elements due to the environmental conditions and the preferences of the user. A small amount of material supply is needed and the structure becomes more economic and sustainable.

The way an Auto-Generated or Bio-Generated dwelling is developed could be easily parallelized with the way living organisms are evolving in Nature. This could have a multi-scalar application; from the tree to the forest, or from the house to the city.

Apart from the constructing development, natural processes could be used for the building's functions as long as defining the ways the production of necessary energy will be done. Building structures could be better configured by imitating or by borrowing characteristics from the natural ones.

MACROTERMES HOUSE

SHICK LEE, Jun
JUNG SOO, Kim

Cornell University

jl649@cornell.edu

People of today´s society has not yet overcome the 20 year lifespan of houses and is repeating the process of building and breaking down. The flood tide of individualism and materialism has caused the increase in use of more expensive and better materials. Due to this, private spaces are becoming more private and the city encloses itself with artificial concrete everywhere. As individualism flows through the roads, the doors, the living rooms and through each of our rooms, spaces that are called public are diminishing and getting out of control. People´s individualism has made the public space an indifferent space and with the loss of management multiple dwelling house is becoming a slum that decreases the value of housing. To prevent buildings from becoming individual and slum, but more sustainable, architecture should not invade to the environment, on the contrary should use it and finally become a part of it. As time goes by, public spaces become the forests, the ponds; they are flexible enough to change according to each unit´s taste. Each of the public spaces suggested functions such as the perspiration holes and the blood vessels of the whole organic body.

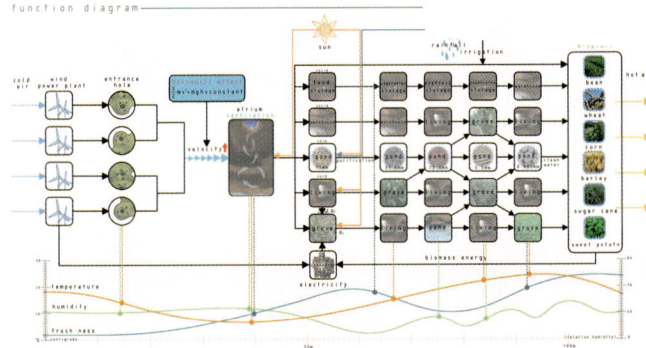

function diagram

DO IT YOURSELF

AUTO- & BIOGENERATED

MATERIALS & CONSTITUTION

RECYCLED & READAPTED

MODULATED & PRE-ARRANGED

DIGITAL FAB

b i o p l a n t   d i a g r a m
s e l f – s u f f i c i e n t   a r e a   e q u a t i o n

required number of persons **x** annual consumption per 1 person / annual production per 1 a = **self-sufficient** arable land (a)

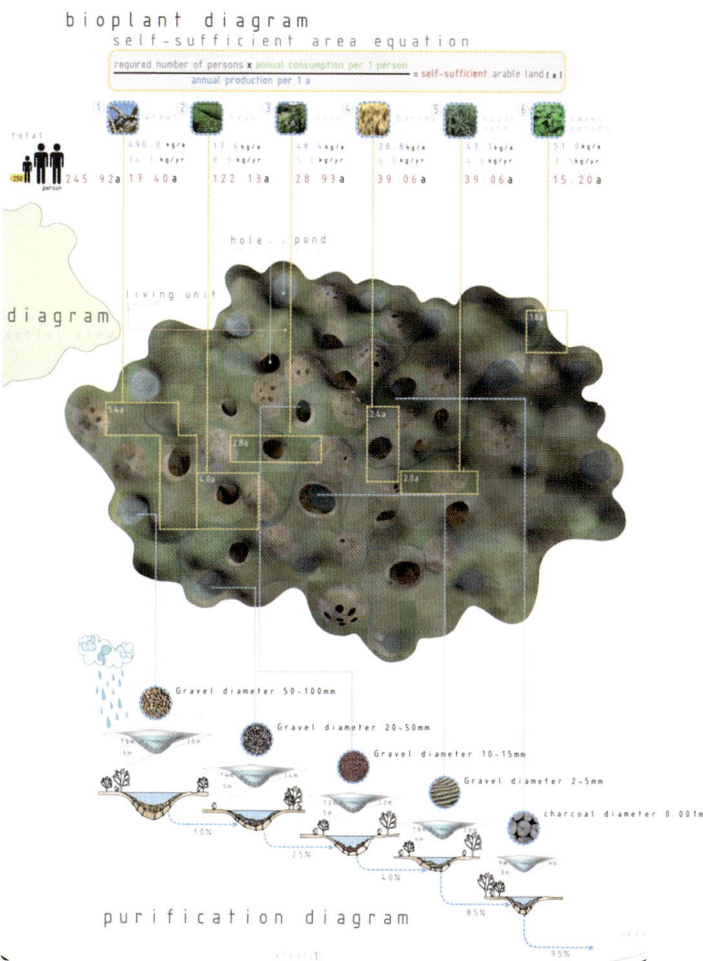

hole.. pond

living unit

diagram

purification diagram

Gravel diameter 50-100mm

Gravel diameter 20-50mm

Gravel diameter 10-15mm

Gravel diameter 2-5mm

charcoal diameter 0.001mm

plan
section

DO IT YOURSELF

AUTO- & BIOGENERATED

MATERIALS & CONSTITUTION

RECYCLED & READAPTED

MODULATED & PRE-ARRANGED

DIGITAL FAB

LANGENBERG, Erno
CORDEIRO, Isabel

ELstudio

info@elstudio.nl

This proposal aims to explore the potential of water surfaces for flexible urban development. The project consists of a pier landscape connected to the main land, it is accessible both by car and by boat. The pier network is support for the prefabricated housing modules, which are plugged into the connection points. The water allows the easy transportation and the mobility of the floating housing units. This means that in time units can be relocated changing the configuration and density of the plan. Also, program such as parking, retail, public areas, etc. can be added or removed according to the needs of each neighbourhood. This program can be placed in floating unit similar to the housing units.

Parametric Design

The pier network is generated by a computer script, assuring the circulation of the floating traffic in present and future, this is important in order to place and remove the prefabricated units. Because of this parametric approach, which defines the basic rules of development, growth becomes flexible: it can occur in different times and places, developed by different parties, at different locations. This adaptability makes the floating-city more sustainable then average sedentary cities since it can in/decrease as well as recycle its program.

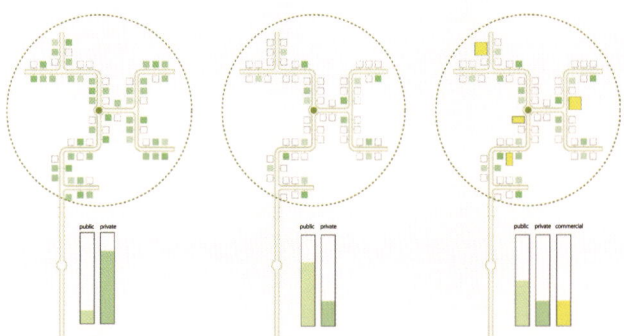

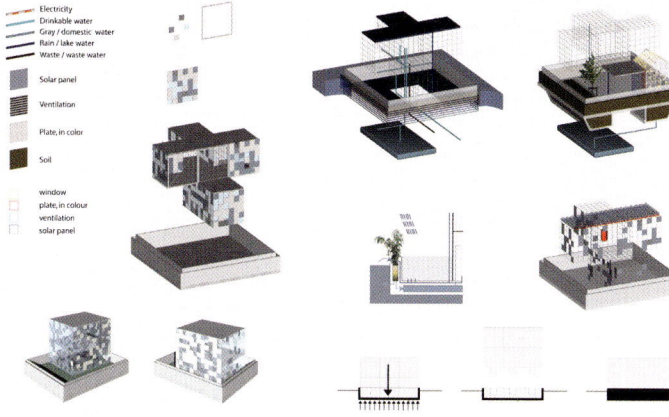

Electricity
Drinkable water
Gray / domestic water
Rain / lake water
Waste / waste water

Solar panel

Ventilation

Plate, in color

Soil

window
plate, in colour
ventilation
solar panel

DO IT YOURSELF

AUTO- & BIOGENERATED

MATERIALS & CONSTITUTION

RECYCLED & READAPTED

MODULATED & PRE-ARRANGED

DIGITAL FAB

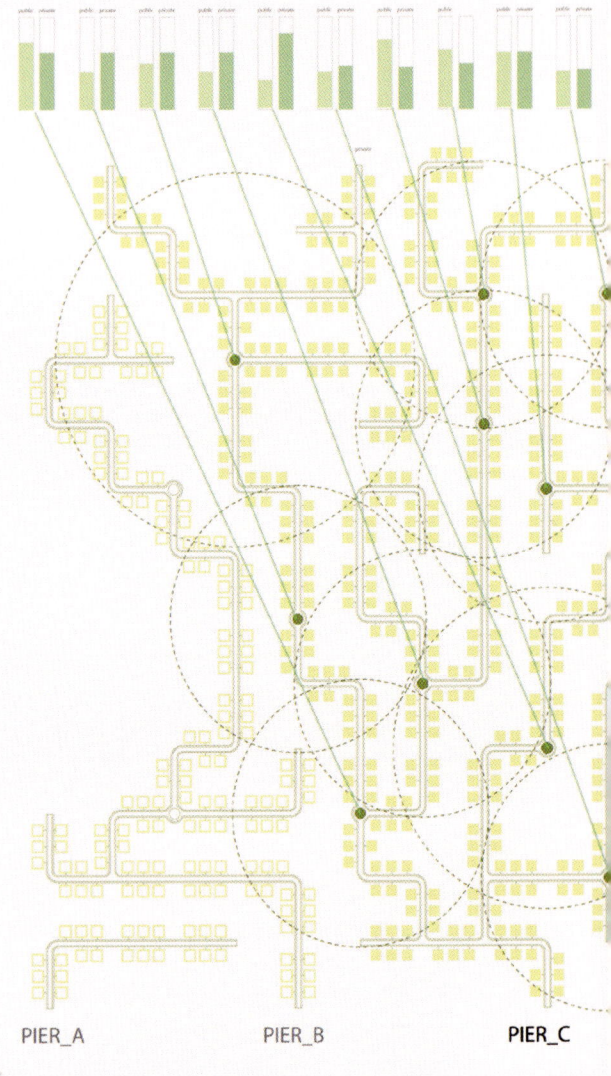

PIER_A   PIER_B   **PIER_C**

500 m

30% window
20% solar panel
8% ventilation
42% metal

40% window
30% solar panel
5% ventilation
25% woodl

45% window
30% solar panel
5% ventilation
20% wood

PIER_D

PIER_E

DO IT YOURSELF

AUTO- & BIOGENERATED

MATERIALS & CONSTITUTION

RECYCLED & READAPTED

MODULATED & PRE-ARRANGED

DIGITAL FAB

ARROYO PAGAN, Edgardo

untitled.psd

untitledpsd@gmail.com

The world has perceived the great limitation of resources and tries to be transformed in order to achieve self-sufficiency. To be self-sufficient cannot be accomplished, for intance, by just building a "green" house made in a secluded beach. Nonetheless, it thrives and becomes a reality when the self-sufficient portrait is applied in a city which in return becomes a sustainable one, in every aspect of its existence, calls for the construction of urban ecosystems and promotes a futuristic society.

The project, is an urban one made out of 25 buildings connected by parks and transit systems, which at the same time function as primary sources of energy and resources. Its morphology proposes an autonomous mobility of the citizen, where the pedestrian, the cyclist and public transportation serve as erasers of the automobilistic dependency.

Buildings are designed in order to provide 100% of their electrical needs. A corrugated metal base overlapped by flexible photovoltaic cells would serve as the main tectonic element of the facades, providing much more solar recollection area than a traditional roof panel system. Water is collected on specially designed roof systems and due to its innate weight, it is transported to each floor and subsequently, to each housing unit.

Unusable dark water would in return be transported to a system of water renovation and compost creation. This creates a cycle of water renovation for the housing units and their respective gardens.

Every material used during construction is left in the site, thus the waste is eliminated and energy displacement is embodied during the project completion.

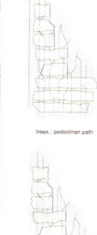

main / pedestrian path

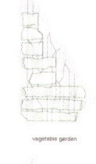

vegetable garden

building / services

green areas / recycle park

DO IT YOURSELF

AUTO- & BIOGENERATED

MATERIALS & CONSTITUTION

RECYCLED & READAPTED

MODULATED & PRE-ARRANGED

DIGITAL FAB

SELF-SUFFICIENT HOUSING

STANKEVIC, Jezi
SPILEVSKI, Romuald

Vilnius Academy of Fine Arts
Warsaw University of Technology

jezi.stankevic@gmail.com

The main purpose of our project is to create a self-sufficient housing using the polluting gases, CO, CO_2 and sunshine. We suggest that energy can be acquired through the reduction of polluted gases using artificial bio-chemical processes based on photosynthesis and greenhouse effect. Homes can be located in areas which are harmful for the human health: areas polluted by CO_2 and CO - roads, tunnels, skyways, crossroads, other planets containing CO_2 in their atmosphere: Mars, Venus.

Our project is a commune of self-sufficient residences which are like artificial trees in the city – they inhale polluted air and exhale clean one. Buildings are armed with "multilayer cell skin" which nourishes the house and its inhabitants. The Bio-reactive cell is integrated into the skin of the building and consists of three parts:

1. external isolation cell filled with CO_2 – protects the inside cells and heats the interior on the basis of greenhouse effect.

2. photosynthesis cell filled with algae which absorbs cell filling gases (containing CO_2) and procreate transforming into biomass which is then (through burning reactor) transformed into electrical energy and heat for the water used at home.

3. thin glass layer filled with color gas, its density can be regulated (from mute to total maximum) in order to change the volume of sunshine coming into the interior. The sunlight is transformed into electric energy by the glass sheets with spinach protein placed fragmentary on "the skin".

The shape of the building is determined by the computer program by providing essential data specific, for each location, which help the optimization of the construction, the position of sun batteries etc.

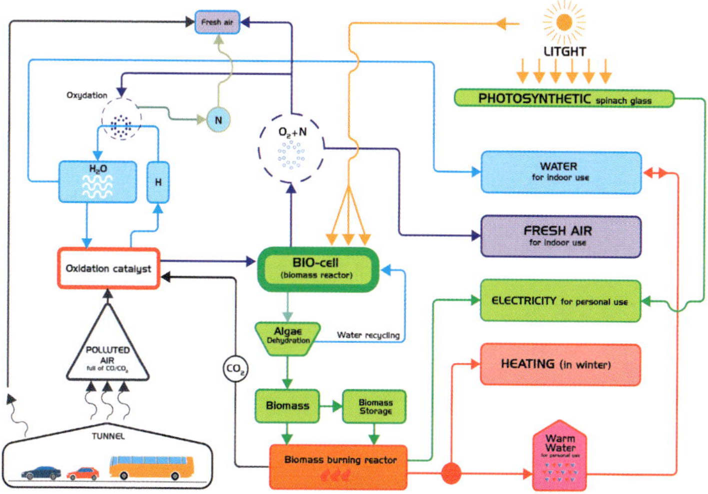

Diagram labels:

Fresh air

Oxydation

N

$O_2 + N$

H_2O

H

Oxidation catalyst

BIO-cell (biomass reactor)

POLLUTED AIR full of CO/CO_2

CO_2

Algae Dehydration

Water recycling

TUNNEL

Biomass

Biomass Storage

Biomass burning reactor

LIGHT

PHOTOSYNTHETIC spinach glass

WATER for indoor use

FRESH AIR for indoor use

ELECTRICITY for personal use

HEATING (in winter)

Warm Water for personal use

DO IT YOURSELF

AUTO- & BIOGENERATED

MATERIALS & CONSTITUTION

RECYCLED & READAPTED

MODULATED & PRE-ARRANGED

DIGITAL FAB

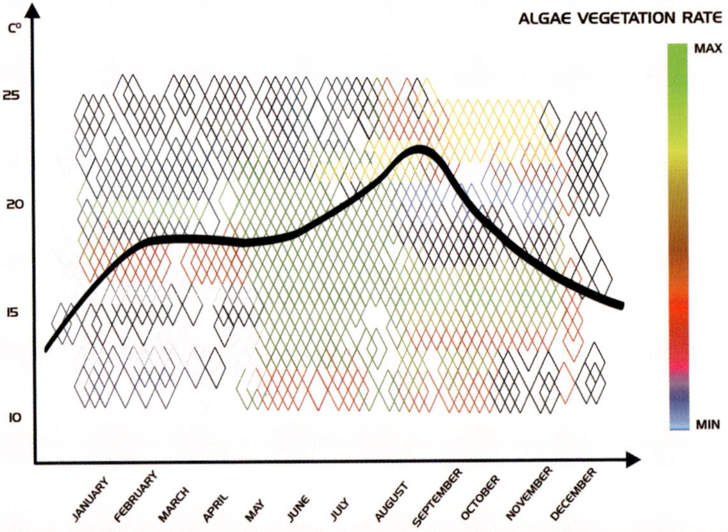
ALGAE VEGETATION RATE
MAX
MIN
C°
25
20
15
10
JANUARY
FEBRUARY
MARCH
APRIL
MAY
JUNE
JULY
AUGUST
SEPTEMBER
OCTOBER
NOVEMBER
DECEMBER

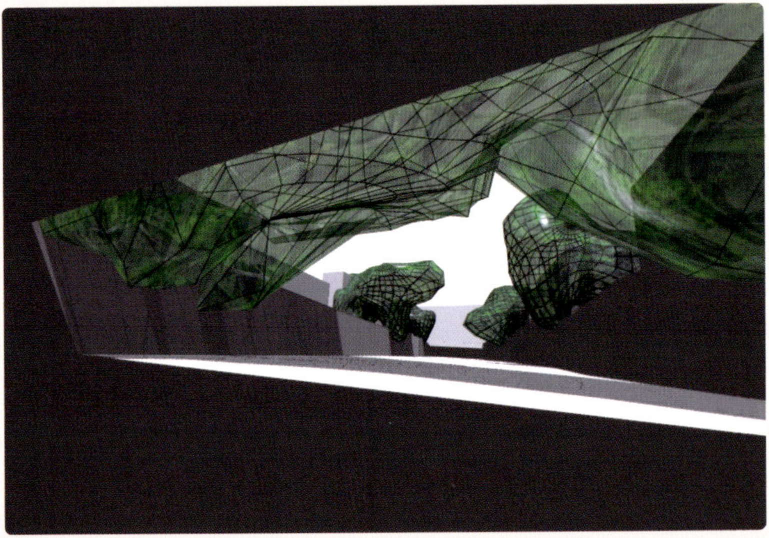

DO IT YOURSELF

AUTO- & BIOGENERATED

MATERIALS & CONSTITUTION

RECYCLED & READAPTED

MODULATED & PRE-ARRANGED

DIGITAL FAB

RISING HOUSE

SCARPINATO, Marco
GUAGLIARDO, Vicenzo
PIERRO, Lucia
MAJORANA, Lorenza
VITRANO, Carmelo

Autonome Forme

info@autonomeforme.it

The rising house is a self-built house, a dynamic device able to build a living space automatically. The rising house recoups "mechanic" aspects which have been excluded from the digital development that has colonised the contemporary imagination. The mechanical device distinguishes what it is from what it becomes, what is taken from history and what belongs to the present. The retrieval of the mechanical form should have an important impact on sustainability as repairs can be made enabling a long life at a low cost.

Concept

The rising house combines advanced technology of energy supply and simple construction using the pantograph mechanism taken from industrial lifting plates. Through the movement of the pantograph the space is configured into one or two levels using the energy stored in the battery platforms. The photovoltaic panels on the roof absorb daylight. The idea of selfbuilding combines prefabrication with automatic construction, building the space without external help.

Autoconstruction / self fab

The ideal sites for the location of the rising house are places with restricted access, both from an energy point of view and the supply of materials. The house can be also used in disaster zones (war, flooding, hurricanes, etc.) which are deprived of infrastructure in terms of energy and organisation.

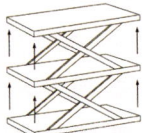
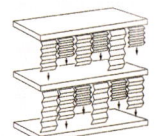

DO IT YOURSELF

AUTO- ¿ BIOGENERATED

MATERIALS ¿ CONSTITUTION

RECYCLED ¿ READAPTED

MODULATED ¿ PRE-ARRANGED

DIGITAL FAB

1ST. ACCUMULATION OF ENERGY IN THE BATTERY PLATFORMS   2ND. START OF THE RAISING PROCESS   3RD. COMPLETION

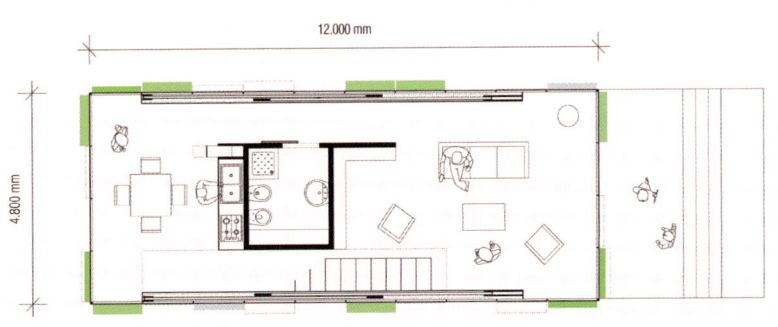

12.000 mm

4.800 mm

Ground floor. 1:100

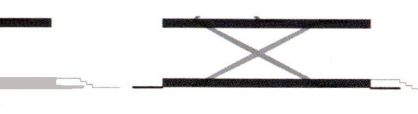

FLOOR   4ST. COMPLETION OF THE FIRST FLOOR   5ST. UNFOLDING OF THE SKIN

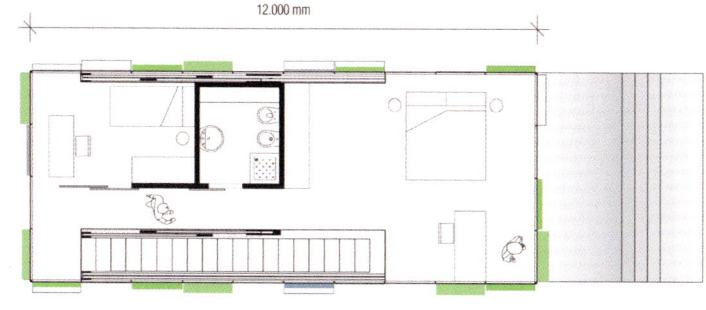

12.000 mm

First floor. 1:100

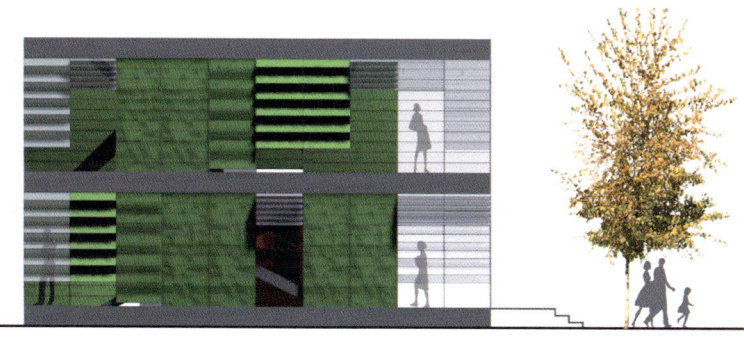

DO IT YOURSELF

AUTO - & BIOGENERATED

MATERIALS & CONSTITUTION

RECYCLED & READAPTED

MODULATED & PRE-ARRANGED

DIGITAL FAB

8F97CD

GROW YOUR HOUSE

Serbia & Montenegro

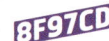

DRASKOVIC, Hana

Arhitektonski Fakultet Univerziteta u Beogradu

hanadraskovic@yahoo.com

The idea is to simply let nature take its course. Forget about imitating, quoting, referring to, etc.

This is a self growing organism that not only lives but designs itself (depending on the location and season) over a recycled steel skeleton. So, leave it grow, water it from time to time and enjoy!

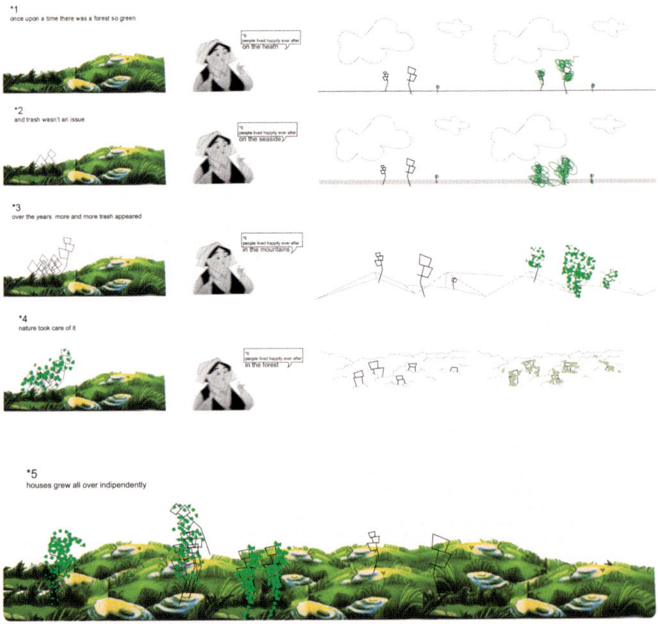
*1
once upon a time there was a forest so green
people lived happily ever after
on the heath
*2
and trash wasn't an issue
people lived happily ever after
on the seaside
*3
over the years more and more trash appeared
people lived happily ever after
in the mountains
*4
nature took care of it
people lived happily ever after
in the forest
*5
houses grew all over indipendently

	January	February	March	April	May	Jun	July	August	September	October	November	December
Fractals												
Sequence												
Nature												
Climate												
People												
Children												

DO IT YOURSELF

AUTO - & BIOGENERATED

MATERIALS & CONSTITUTION

RECYCLED & READAPTED

MODULATED & PRE-ARRANGED

DIGITAL FAB

PINEIRO GARCIA, Ana

Architect

ana_pineiro@coac.net

The idea is to build a self sufficient net with a changeable number of dwellings in order to give solutions in cases such as temporary occupations, emergency or unstable circumstances. First, the project tries to comprehend the essential characteristics of the territory: water behaviour; and at the same time, continuous human actions that transform it. The superposition of stripes in the inhabited fabric is based on the continuous elements that are mostly crossing. Permeability stripes are in relation to water corridors and sand roads.

Grouping law sector development > group of cells > cell = minimum unit

The structure will be built at the period of its necessity and according to this grouping law, it will give rise to its own form once penetrated by light. Light and air are considered as materials that build architecture.

Build, function and presence

Its conception in the form of a mecano allows a vertiginous construction on big surfaces. The components of the system were characterized by their lightness. They could be raised easily by one or two non-qualified workers. Its conception, similar to the American Balloon Frame, a reticule of light elements with small lights, let it being used everywhere. The result is a delicate structure, adapted to the natural topography. A skeleton of sustaining elements working in a strict balance.

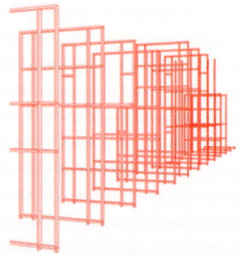
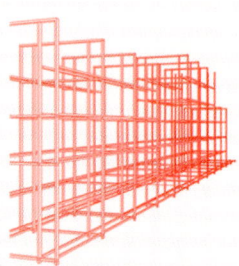

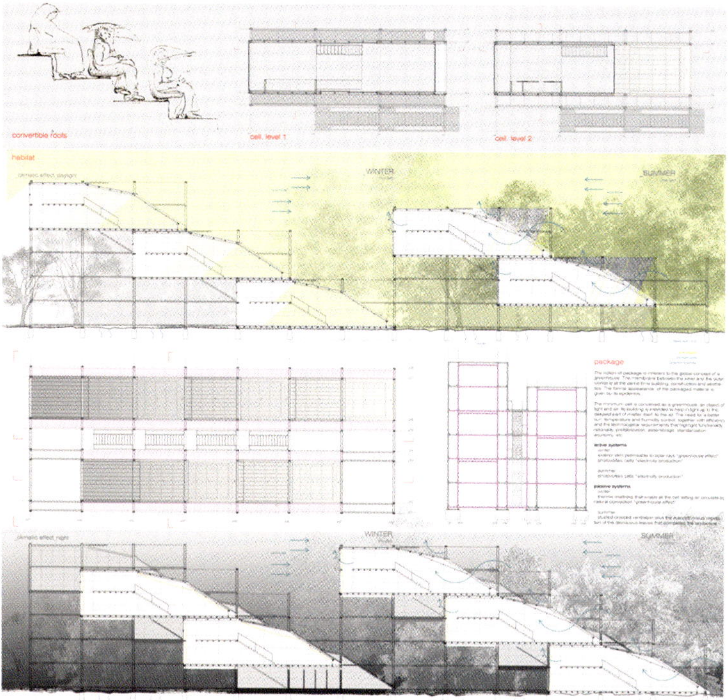

DO IT YOURSELF

AUTO- & BIOGENERATED

MATERIALS & CONSTITUTION

RECYCLED & READAPTED

MODULATED & PRE-ARRANGED

DIGITAL FAB

convertible roofs

cell. level 1

cell. level 2

habitat

climatic effect (daylight)

WINTER

SUMMER

package

climatic effect night

WINTER

SUMMER

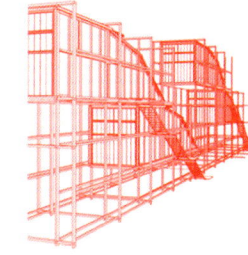

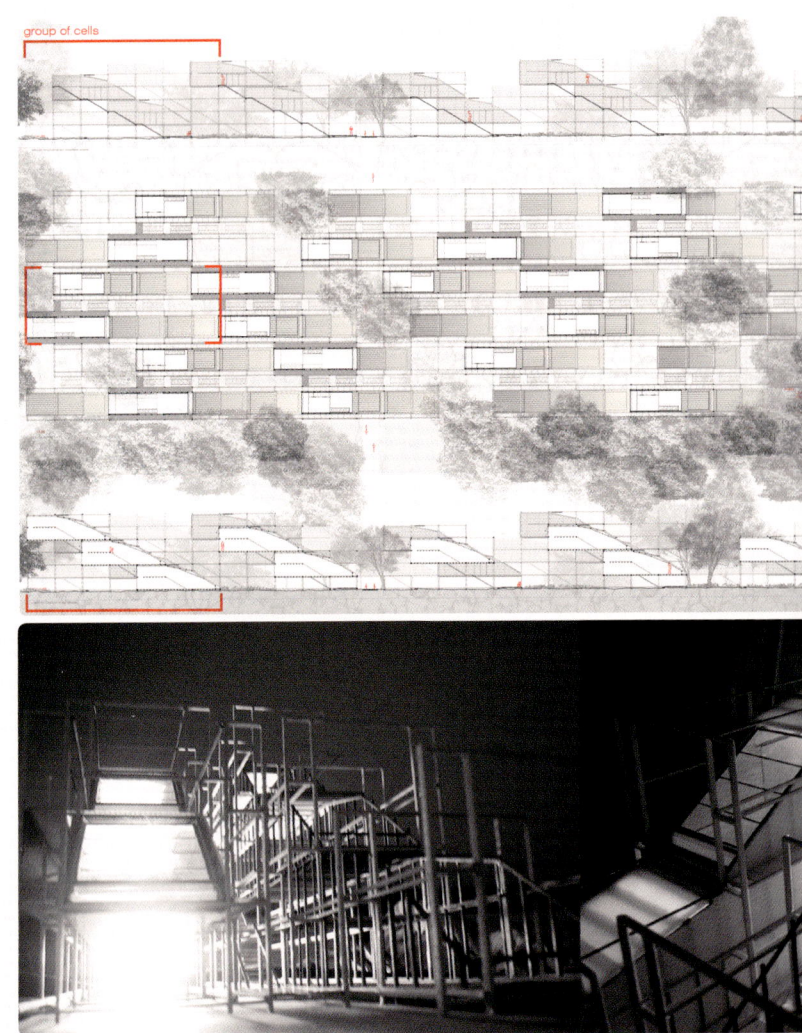

group of cells

DO IT YOURSELF

AUTO- & BIOGENERATED

MATERIALS & CONSTITUTION

RECYCLED & READAPTED

MODULATED & PRE-ARRANGED

DIGITAL FAB

TECNO-ECOLOGIES

Spain

BALLESTEROS, Simón
GARCÍA PÉREZ, Elena

ANTIFABRIC.Interdisciplinary Lab

ivan@antifabric.com

Ecology is the study of the relationship between organisms and their environment. This definition also suits the discipline of architecture surprisingly well: in our view one of the central tasks of architecture is to provide opportunities for habitation through specific material and energetic interventions in the physical environment.

Techno-ecologies explore behaviours and logics in nature and the technological implementation as mechanism to generate a self-sufficient housing. The proposal studies the cactus morphology as organism able to resist extreme weather condition as an autonomous and self-sufficient entity.

Also the project distinguishes two levels of information: the general conditions (geometry, behaviour, energy) and the local adaptations in terms of environmental parameters (lighting, thermal radiation, insulation, wind). The interaction of both levels of information establishes the strategy of the proposal, generating a set of possible variations.

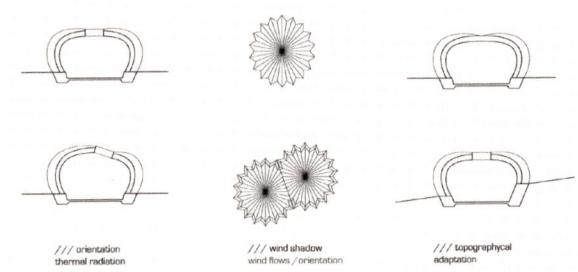

/// orientation thermal radiation

/// wind shadow wind flows / orientation

/// topographycal adaptation

DO IT YOURSELF

AUTO- $ BIOGENERATED

MATERIALS $ CONSTITUTION

RECYCLED $ READAPTED

MODULATED $ PRE-ARRANGED

DIGITAL FAB

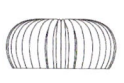

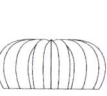

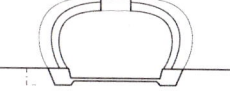

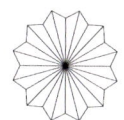
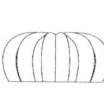

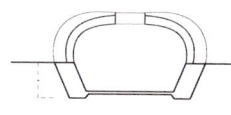

/// earth insulation
Different range of
excavation accordign to
the local topography
and insulation level.

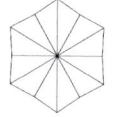
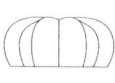
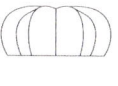

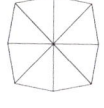

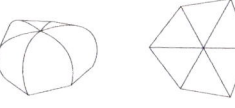

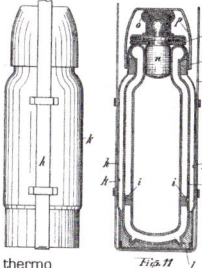

thermo   Fig.11

DO IT YOURSELF

AUTO- & BIOGENERATED

MATERIALS & CONSTITUTION

RECYCLED & READAPTED

MODULATED & PRE-ARRANGED

DIGITAL FAB

AMARA, Andrew

Makerere University

andre_andre44@yahoo.com

On the Outskirts of Kampala city, lies Kinawata slums - a spread of shacks, shanties, shelters and a few brick houses. The Unique characteristics about this site are:

— Intimate organic density of the units that create inbetween YARDS

— the YARD is a 3rd layer of connectivity

— the YARD is a functional space, enjoyed by all age groups.

The units have developed organically, depending on the needs of the residents - from housing a new family member in a new room to set up the shelter for a small business.

To facilitate this vibrant community its dynamic livelihoods, the design proposes a house that can grow/ evolve, according to the dwellers' needs. To aid this organic growth, data/info is imprinted on to the house enclosures. The dweller can read and interpret this data, and then can be able to modify the house. This DWELLER - HOUSE interaction turns the house into a living building that can evolve and change.

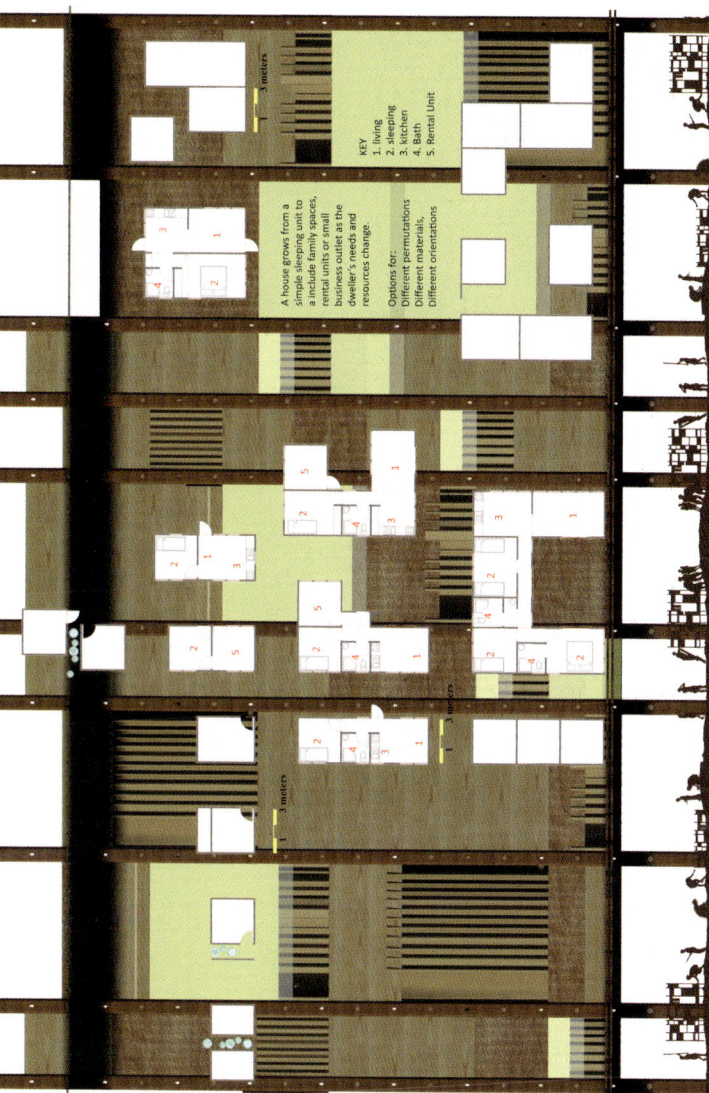

KEY

1. living
2. sleeping
3. kitchen
4. Bath
5. Rental Unit

A house grows from a simple sleeping unit to a include family spaces, rental units or small business outlet as the dweller's needs and resources change.

Options for:
Different permutations
Different materials,
Different orientations

5 meters

3 meters

3 meters

DO IT YOURSELF

AUTO- & BIOGENERATED

MATERIALS & CONSTITUTION

RECYCLED & READAPTED

MODULATED & PRE-ARRANGED

DIGITAL FAB

KERIMOL, Levent

Architectural Association

lev100@gmail.com

Rather than an experimental prototype that may not prove popular in the housing market, this project proposes a realistic mechanism and process for the creation of a large-scale suburban development in which houses are built or developed by the residents themselves, as opposed to developers or architects. A consideration of the social, political, and economic conditions for the adoption of self build as a mainstream alternative, and the implications such changes might have, suggests that independence from volume housebuilders and large energy corporations must be combined with local collaboration and negotiation. To briefly summarise the structure; the land development trust is initially a partnership to create a housing related social enterprise. Residents become part of the trust as they buy plots, and have a say in how it is run. The trust has a long-term stake in the area, as it retains the freehold of the land, and collects annual ground rents, which increase as the area matures. These go towards maintaining roads and trees, and running local initiatives such as car pools.

Smaller groups of residents manage their respective commonhold gardens and their boundaries. The right to develop a plot comes with certain legal covenants intended to prevent amateur developers from build-to-let or build-to-sell, and encourage unique houses for long-term residents. However there are no restrictions on the style or method each household uses to build. The existing building methods available to households have been analysed. Certain organisational innovations in the construction industry could bring together some of the practical advantages of prefabrication, and streamlined management, without losing the uniqueness and individuality possible with self-build. A software tool (scripted as part of this project) generates a range of options based on the household's individual profile. Architects can advise on best choice. This method could be semi-automated to direct any digitally custom fabricated system, such as structural insulated panels, offering speed, flexibility and impressive thermal properties. Another piece of software allows residents to be part of making the local layout by claiming and sharing their own plots. Residents are obliged to interact with each other from the outset, in the hope of limiting social atomisation and selfishness. Urban detailing addresses the architectural qualities of these moments of negotiation.

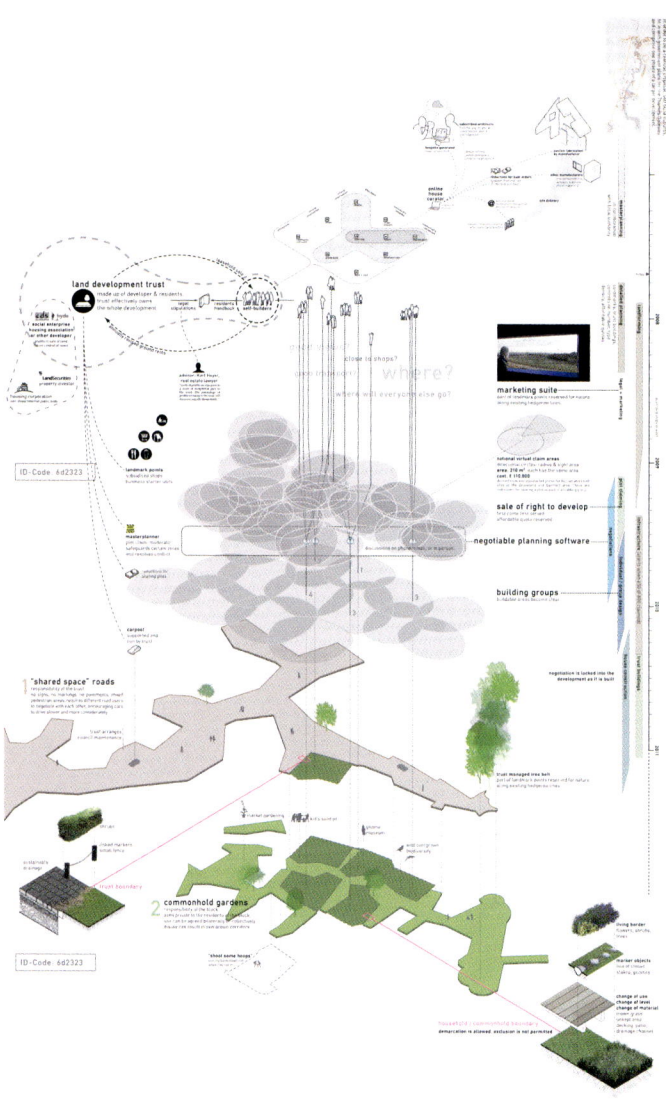

DO IT YOURSELF

AUTO- & BIOGENERATED

MATERIALS & CONSTITUTION

RECYCLED & READAPTED

MODULATED & PRE-ARRANGED

DIGITAL FAB

land development trust
made up of developer & residents
trust effectively owns
the whole development

marketing suite

sale of right to develop

negotiable planning software

building groups

1 "shared space" roads

2 commonhold gardens

ID-Code: 6d2323

ID-Code: 6d2323

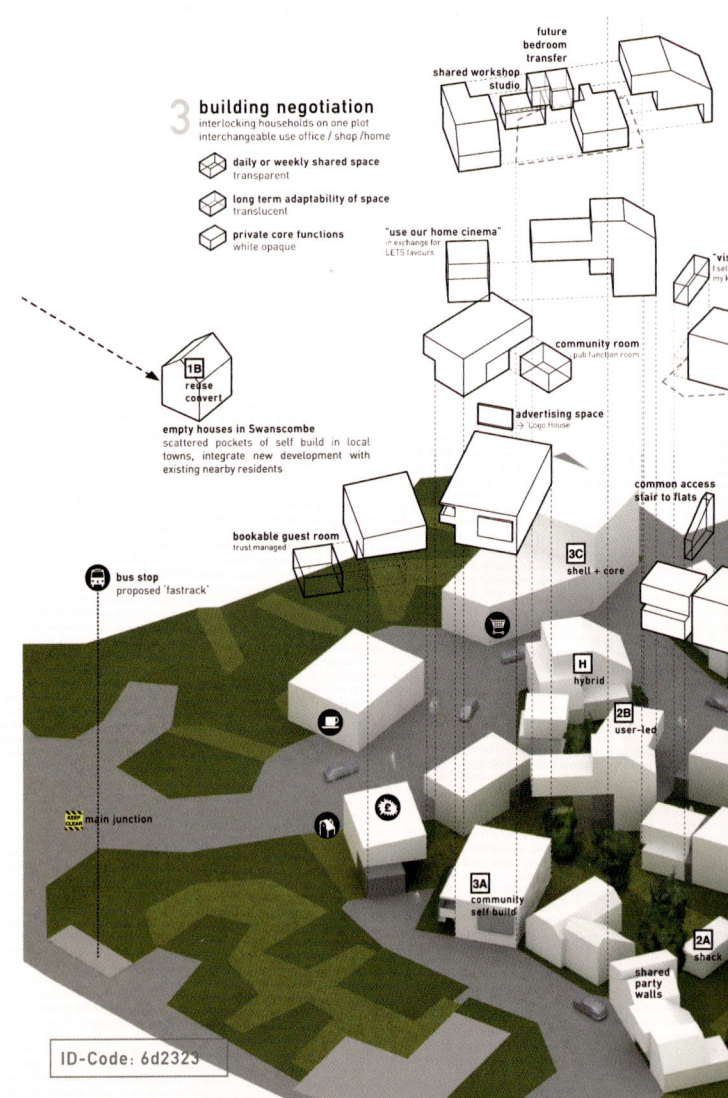

3 building negotiation
interlocking households on one plot
interchangeable use office / shop /home

⬡ **daily or weekly shared space**
transparent

⬡ **long term adaptability of space**
translucent

⬡ **private core functions**
white opaque

future
bedroom
transfer

shared workshop
studio

"use our home cinema"
in exchange for
LETS favours

"vis
I sell
my ki

community room
pub function room

1B
reuse
convert

empty houses in Swanscombe
scattered pockets of self build in local
towns, integrate new development with
existing nearby residents

advertising space
→ 'Logo House'

common access
stair to flats

bookable guest room
trust managed

bus stop
proposed 'fastrack'

3C
shell + core

main junction

H
hybrid

2B
user-led

3A
community
self build

2A
shack

shared
party
walls

ID-Code: 6d2323

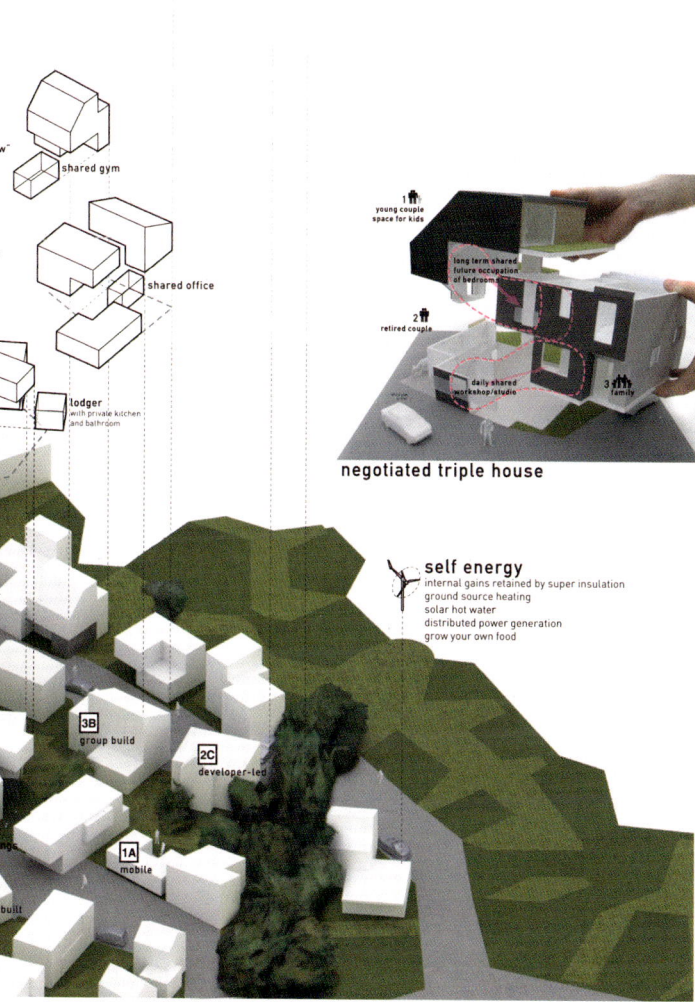

ow¯

shared gym

shared office

lodger
with private kitchen
and bathroom

1 👫
young couple
space for kids

long term shared
future occupation
of bedrooms

2 👫
retired couple

daily shared
workshop/studio

3 👪
family

negotiated triple house

self energy
internal gains retained by super insulation
ground source heating
solar hot water
distributed power generation
grow your own food

3B
group build

2C
developer-led

inge

1A
mobile

n built

DO IT YOURSELF

AUTO- & BIOGENERATED

MATERIALS & CONSTITUTION

RECYCLED & READAPTED

MODULATED & PRE-ARRANGED

DIGITAL FAB

DAMJANOV, Nikola
KRSMANOVIC, Slobodan
BOJOVIC, Petar

Faculty Of Architecture/ Electrical
Engineering/ Mechanical Engineering of
Belgrade

damjanmx@yahoo.com

We suggest that existing floating houses are replaced with short lived
ephemeral structures, that are mainly built out of recycled elements which
are produced on the island. The users assemble their own new floating houses, on one of the potential positions, out of the elements that are provided
by the community center / "Mothership". All elements, beside solar panels
and inflatable pontoons have very limited expiration date. The platform and
wall panels are made out of mixture of recycled paper and organic glue, and
are 5 or 15 cm thick. After they are installed and exposed to moisture, they
are usable for a period of approximately 7 days. Expiration date of proposed
floating house, matches the average time, its users spend on the island during a single visit (2 - 10 days). All the things that are necessary for the floating
house to be assembled as long as the ones needed so as to live there for a
short period of time are prepared, produced and recycled in the "Mothership". The production of electrical energy is realized in 3 possible ways:

— harvesting the river stream power

— using the solar energy

— recycling biological waste

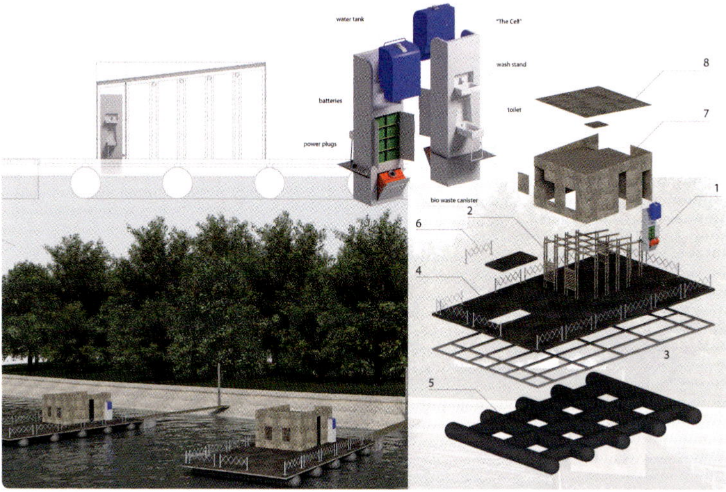

water tank

"The Cell"

wash stand

batteries

toilet

power plugs

bio waste canister

8

7

1

2

6

4

3

5

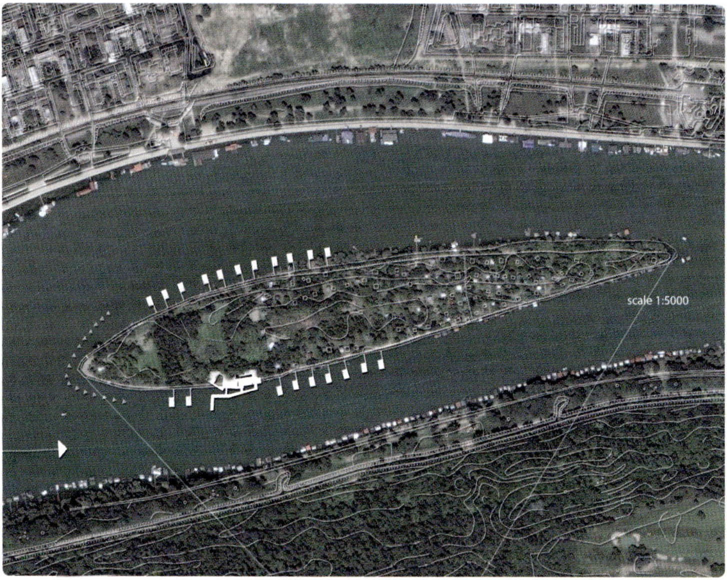

scale 1:5000

DO IT YOURSELF

AUTO- & BIOGENERATED

MATERIALS & CONSTITUTION

RECYCLED & READAPTED

MODULATED & PRE-ARRANGED

DIGITAL FAB

HANDY, Katie
PERERA, Dinesh

UC Berkeley

handy.katie@gmail.com

24.5 Million members of the the global population fall under the category of IDP — internally displaced peoples. This disturbing figure is only a fraction of those who do not have access to the the most basic life-sustaining amenities. The dispair caused by these conditions often fuel ongoing violence, as is the case in Sri Lanka, which has been locked in a bloody civil war for the last 24 years, resulting in 600,000 conflict IDPs in the previous year alone. The local infrastructure, weakened by the war, has been unable to provide or distribute a permanent housing solution for IDPs or calm political tensions. Providing access to global systems as well as natural resources will allow communities to grow independently from the state.

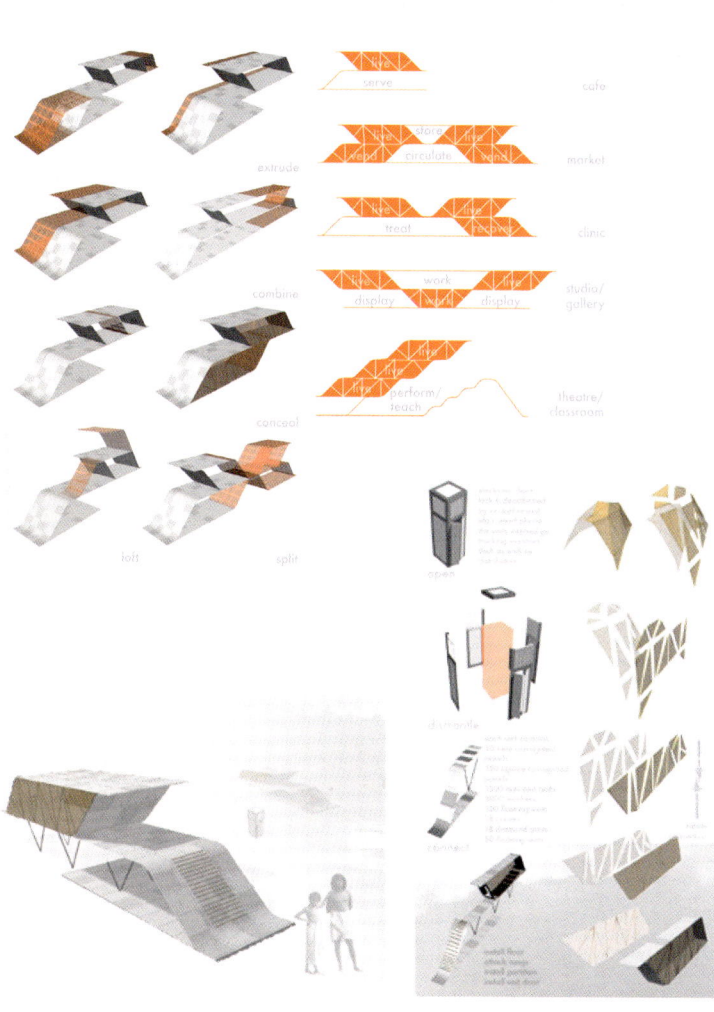

extrude

combine

conceal

loft

split

live
serve
cafe

live store live
vend circulate vend
market

live live
treat receive
clinic

live work live
display work display
studio/
gallery

live
live perform/
teach
theatre/
classroom

open

dismantle

connect

DO IT YOURSELF

AUTO- & BIOGENERATED

MATERIALS & CONSTITUTION

RECYCLED & READAPTED

MODULATED & PRE-ARRANGED

DIGITAL FAB

PIERCE, Jason
BOSSERT, Bennett
HEADLEY, Dustin
LAI, Tzu-Ching
MARINARO, Kathryn

PERRY, Kyle
POORE, Tom
VERMILLION, Joshua
WHITE, Kaitlin
YESHAYAHU, Shai

Southern Illinois University

piercel@siu.edu

[COLLECTIVE KNOWLEDGE] [networking] [sharing] [library] [p2p]

Our dwelling design connects with others locally, regionally, and globally through a peer-to-peer network to share and obtain information needed to optimize its configuration. This could happen automatically, and may need no attention from the inhabitant. Houses become aware of themselves in relation to the world and to other structures via GPS, which can also aid in urban planning analysis. Buildings collaborate to form whole neighborhoods, or even cities that create effective vehicle/pedestrian circulation, quality outdoor space, and community environments. And structures will alert others in different areas of the world of major environmental changes, thereby allowing enough time to optimize before extreme conditions reach those dwellings.

[NANOBOT] [background] [bottom up]

Polycarbonate Nanobots are smaller than a blood cell, lighter than air, and have a single microprocessor, which sends electrical pulses. These pulses send signals to the nanotubes, which grab onto another nanobot's nanotubes to join the network and transfer information. They self-replicate by pulling carbon molecules off of carbon dioxide molecules, and creating parts for a new nanobot – a process which happens in a matter of milliseconds. Much like a factory, certain ones would create specific pieces to collectively build a new one. This "factory" would be set up by linking certain series' of nanobots to others, much like DNA.

[CONTROL] [limits] [assembly] [structure]

The user receives what appears to be a solid object made up of trillions of nanobots. The "object" is hooked up wirelessly to a computer where it receives its instructions from the program. Building codes and regulations are set in the database as limits, which mandate some rules. These limits are set by "the regulators" also known as architects, engineers, and scientists to ensure the development of a healthy infrastructure to the city, prevent the nanobots from running out of control, and deter the general greed of the user. Each individual nanobot finds its exact coordinates for that place and time through a GPS. Once the limits are placed and position is located, the box is linked up to the network where its information is fed into the database and the shape of the house is calculated. As the building grows, each nanobot will self-replicate more nanobots

DO IT YOURSELF

AUTO- & BIOGENERATED

MATERIALS & CONSTITUTION

RECYCLED & READAPTED

MODULATED & PRE-ARRANGED

DIGITAL FAB

DAISUKE, Nagatomo
JAN, Minnie

MisoSoupDesign

dnagatomo@misosoupdesign.com

The urban density causes critical sustainability problems, for example, lack of new construction field to cover over the increasing population, demolition of native green field, pavements covered by asphalt, and human waste led to landfill contaminating the ground. In the case of NYC, especially Manhattan, these problems are intense due to the city´s density. The buildings are aligned next to each other, since the land is extremely expensive, what is more in the cases of a building extension or a construction of a new lowrise one. However, those vertical developments create the odd void condition between buildings, due to height difference.

This designing proposal is focusing on this vertical void condition so as to create self-sufficient housing which takes the advantage of the large development of the city. Living structures form in specific ways to achieve maximum strength in order to survive. These formations are minimal, ecological and sustainable. The key to create complex patterns, the system of monogenous, can be found in these formations. The bubble diagram shows a clear transformation of units according to the articulating rule as well as the complexity created by a single unit. Each cell, represents residential units and is organized by means of bubble structure configuration.

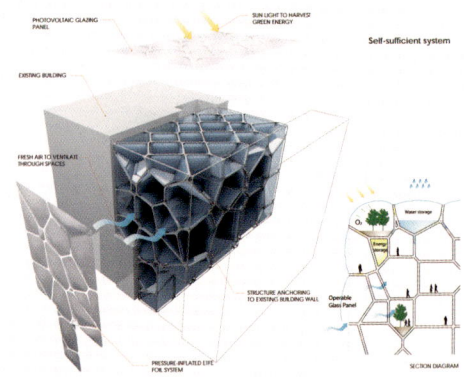

DO IT YOURSELF

AUTO- & BIOGENERATED

MATERIALS & CONSTITUTION

RECYCLED & READAPTED

MODULATED & PRE-ARRANGED

DIGITAL FAB

HAIR SALON & SPA

AVEDA

ARJOL ACEBAL, Ignacio

Gerschfeld, Arjol & Asoc.

nachoarjol@hotmail.com

What to do the American troops in Iraq, Darfur refuqees. New Orleane evacuees, people from Pisco, Peru and many other people living in extreme poverty conditions have in common?

Wrong political decisions, natural disasters and the lack of resources and planning are factors that usually generate massive displacement of populations in short periods of time. This creates temporary settlements with awful life conditions due to the lack ot proper sanitation, sewage systems and access to potable water and energy.

Self-Sufficient Sanitary Unit

Social Objective: to improve conditions of life in temporary settlement. Environmental Objective: to stop contamination from human waste; ensuring the efficient use of potable water; to generate electric energy and natural gas. Economic Objective: providing a low cost sanitary solution and reducing the energy waste.

Usage

The unit could be used on the development of new houses or adapted to existing ones

It would be made out of 3 parts: .

1. Top: electric generation, water storage and heating.

2. Body: Bathroom and kitchen.

3. Bottom: Biodigester.

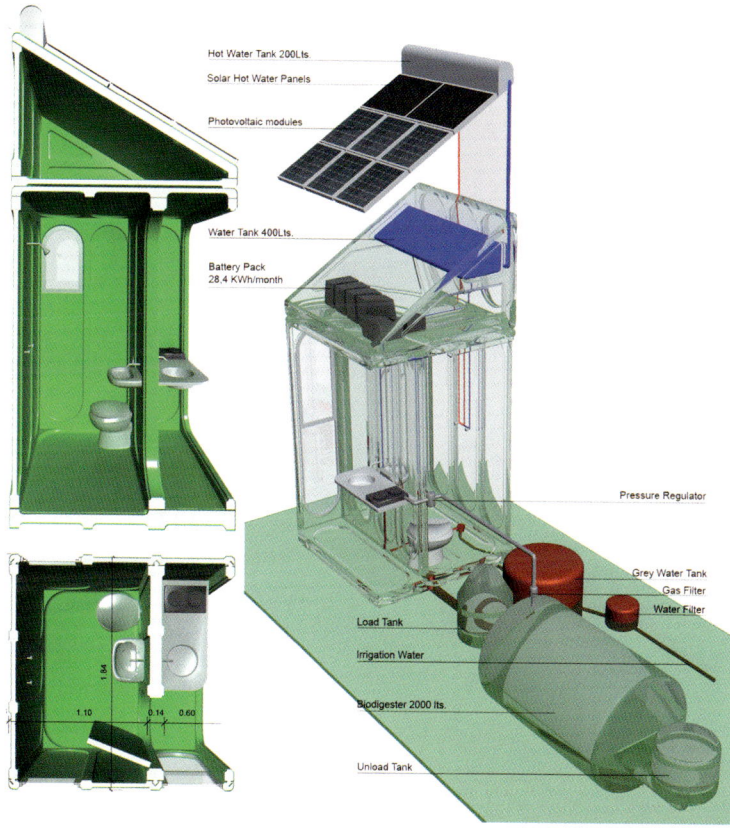

Hot Water Tank 200Lts.

Solar Hot Water Panels

Photovoltaic modules

Water Tank 400Lts.

Battery Pack
28.4 KWh/month

Pressure Regulator

Grey Water Tank
Gas Filter
Water Filter

Load Tank

Irrigation Water

Biodigester 2000 lts.

Unload Tank

1.84

1.10   0.14   0.60

DO IT YOURSELF

AUTO- € BIOGENERATED

MATERIALS € CONSTITUTION

RECYCLED € READAPTED

MODULATED € PRE-ARRANGED

DIGITAL FAB

XIN, Li

Hunan University

lee84990805@yahoo.com.cn

Build your "agri- architecture", which has a terrace field on it's skin!

— Living in the "agri-architecture", former farmers can plant crops on their own house, which could also be a great sightseeing and picnic location for the citizens living in the skyscrapers. Therefore, beside feeding themselves, farms can also gain extra income from the house.

— Healthy and fresh strawberry prices has risen to 1D-18RMB per unit, which is much higher than rices or wheat, therefore, the project proposes a specific planting strategy to be performed.

This self-sufficient system uses natural energy. The system consists of a hand-work-solarshower, a marsh-gas-system, which gives out from human and animal waste, produces clean energy for the family, and a rule for collecting raindrops, which can be used to irrigate crops and clean the toilet.

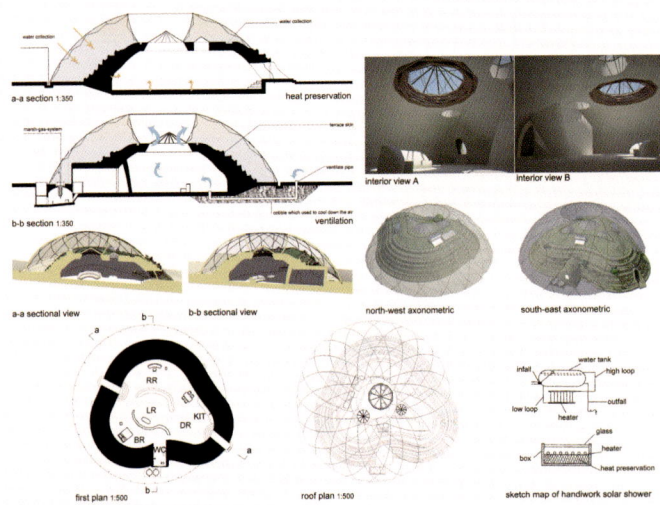

a-a section 1:350   heat preservation

b-b section 1:350   ventilation

a-a sectional view   b-b sectional view   north-west axonometric   south-east axonometric

interior view A   interior view B

first plan 1:500   roof plan 1:500   sketch map of handiwork solar shower

handiwork solar shower
and water tank.

strawberry grow on the
terrace field "wall"

DO IT YOURSELF

AUTO- & BIOGENERATED

MATERIALS & CONSTITUTION

RECYCLED & READAPTED

MODULATED & PRE-ARRANGED

DIGITAL FAB

EL-SOUDANY, Moamen

Architect: Moamen M. El-Soudany

archmsoudany@gmail.com

"Today, industrialists all over the world are reviewing what they have done", this was said by H. Fathy, a great contributor to the modern Egyptian Architecture.

We used the ancient technique of construction methods and materials to obtain the new desert dwelling. The project proposes an arquitectural technique with the use of [EARTH SHELTRED BUILDING], that can promote intergration between the houses and the hard climatic conditions in the deserts. Local trained inhabitants are created to make their own materials and build their own buildings.

[Agricultural reclamation regions]

The areas that were chosen so as to be studied were the ones that meet the characteristics that a region should have so as a new agricultural community to be organized, concerning both self-sufficiency and energy resources. In other terms there are some reclamation deserted lands that suffer from the danger of [Ground MINES] deposits from the [World WAR II]. The concept of [Green Mines community] is suggested to be the symbol of the developing idea. However, there are many samples of these agricultural desert regions in countries like (Saudi Arabia - Egypt - Israel - Lypia) which were studied so as the Urban and Architectural Community Idea "Green Mines community" [G-Mines] to be approached.

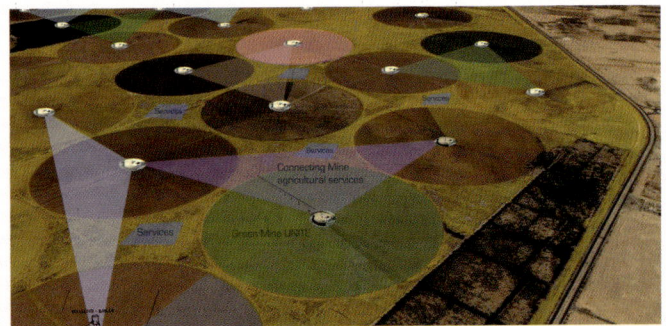

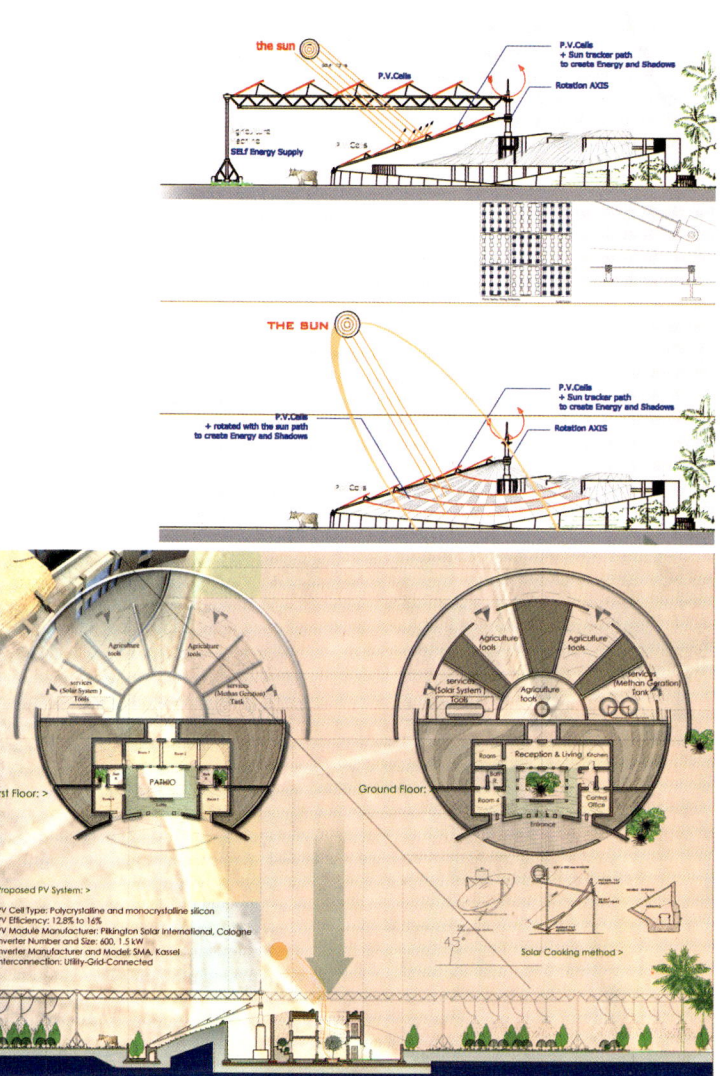

DO IT YOURSELF

AUTO- & BIOGENERATED

MATERIALS & CONSTITUTION

RECYCLED & READAPTED

MODULATED & PRE-ARRANGED

DIGITAL FAB

MATERIALS AND CONSTITUTION

This category tries to explore the two limit approaches for a self-sufficient housing to be created; either the physical one, depending on the materials chosen, or the theoretical one, depending on the constitution and the relationships established.

As far as the physical approach is concerned, there have been used different materials that can give a variety of characteristics in the construction. New materials and processes could be established that could guarantee the self-sufficiency of the building as long as its sustainability. Building skins have been proposed that can easily adapt to different or changing environmental conditions. The structure could finally interact with the ambient that is hosting it. Living organisms, with specific characteristics, could eventually become important structural parts of the building; just the same way traditional housing was constructed.

As said above, apart from the physical approach for a more sustainable society to be formed, there have been proposed and more theoretical ones, mostly by the establishment of new relationships. The building sites and the cities themselves

DO IT YOURSELF

AUTO- & BIOGENERATED

MATERIALS & CONSTITUTION

RECYCLED & READAPTED

MODULATED & PRE-ARRANGED

DIGITAL FAB

could be seen as spaces where collective or individual actions could be performed; as a result the space is transformed and reconfigured. Moreover, the living conditions in an area could become better, only by promoting participation and collaboration between the users-agents. What is more, this idea of establishing new relationships and interconnections, more sustainable ones, resulting in the development of new housing conditions, could be seen also vice versa.

The contemporary way of living is promoting the sharing and the transmission of information. Each structure could be seen as an amount of information which is carried by its elements. This information could be transmitted and shared all over the world. Each one of the structural elements will eventually be transformed due to the needs of the user as long as due to the environmental conditions of the area that it will be placed. This condition of sharing information could finally result in the creation of more economically sustainable and customized housing-products.

SIDDHARTH, N   Srishti School of Art & Design
DUMITRU, Mihaela

                                     sidhrth@gmail.com

The future of self-fabricated and self-sufficient dwellings instantly & simultaneously addresses a twofold issue - that of an evolving collective intelligence (how multiple competing 'agents' co-evolve, given the same base-terrain); and that of a DIY ethic (wherein one is allowed to unfold an extended self into a larger whole, with minimal interruptions/interventions/conflicts (with competing 'others').

x_Nodes are seen as intersection points of this intelligent mesh (Right: mesh evolution) as seen fit/optimal as sites for intervention/construction, minimising conflict/s that arise out of competition for resources.

Once an x_Node is established as a site for construction, the user/ inhabitor initiates &utilizes the mobile/web-interface to select/ generate a desired spatial configuration. The first step is a programmatic pattern/ circuit of possibilities that is imprinted on the x_Node terrain by a swarm of bots. Once this is done, a Material-explorer governed by back-end combinatorial rules helps select desired Grow_cubes (intelligent 3d CA/tangible building block frameworks that can comprise any selected material within) that once placed at appropriate points on the site-circuit, they are transform into desired configurations.

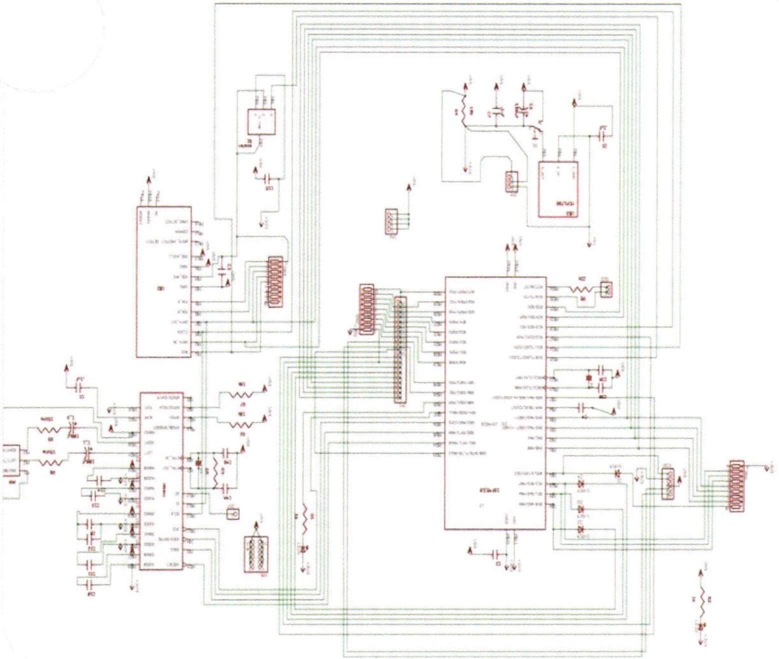

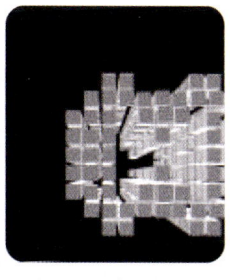
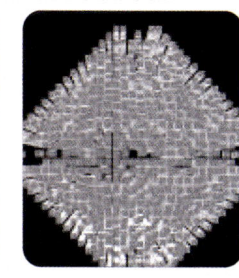

DO IT YOURSELF

AUTO- & BIOGENERATED

MATERIALS & CONSTITUTION

RECYCLED & READAPTED

MODULATED & PRE-ARRANGED

DIGITAL FAB

PENA, Adriana   ITESM Campus Guadalajara
VILLASENOR, Isabel
                                        happygaby101@gmail.com

Location: Sayula desert is located in Mexico, with an extension of 168 km², this once prominent water formation is now dominated by dry, parched land 10 months a year with temperatures oscillating between 35-0 C. Precipitation is present between the months of July and August which provides the dry lagoon with a total of 578.70 mm of water per year. Floodings are common during rain season and can create fleeting water bodies that can reach depths of up to 2mts.

Creating a bond between the natural and the artificial environment, this temporary and self-sufficient housing module uses the dynamics of motion to fully adapt to the ever changing weather conditions.

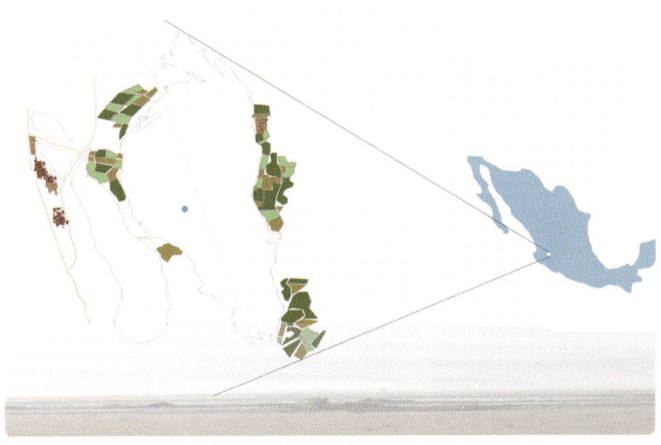

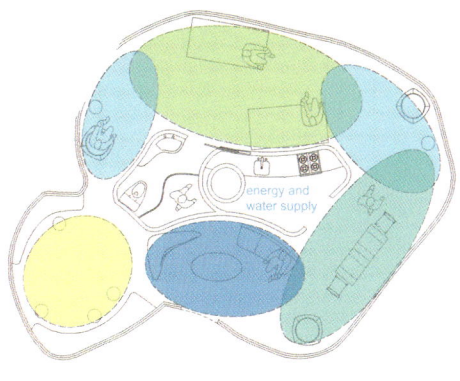

energy and
water supply

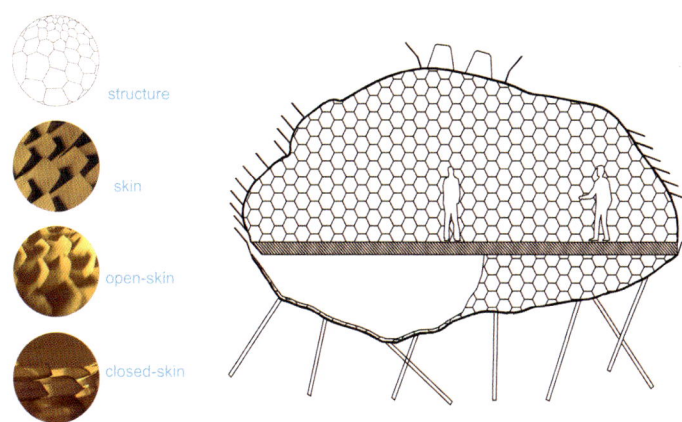

structure

skin

open-skin

closed-skin

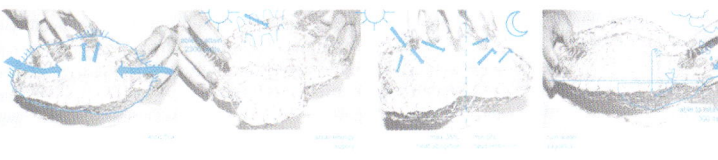

DO IT YOURSELF

AUTO- & BIOGENERATED

MATERIALS & CONSTITUTION

RECYCLED & READAPTED

MODULATED & PRE- ARRANGED

DIGITAL FAB

KANNAN, Mathan
SOUNDARI, Kamala

S.A.P. – School of Architecture
and Planing, Anna University

mathankannank@rediffmail.com

Problem: Housing unit for tsunami rehabilitation.

Site: Coast line of Southern India which has been hit by tsunami.
The proposal is suitable for any costal informal settlements.

Aim: The objective of our design is to use the bamboo as the structural building material which can be planted along the sea coasts. Moreover, we use the bamboos without cutting down them.

Ideas held: eco-friendly design, we use bamboo for construction without removing it from the ground. What is more, the method of construction proposed is cost-effective. The mechanism is not complicated and the labor required should not necessarily be skilled. The interior space can be redefined at any time even after the completion of the structure. The green bamboo leaves acts as heat resisting envelope and enriches the thermal comfort level of interiors.

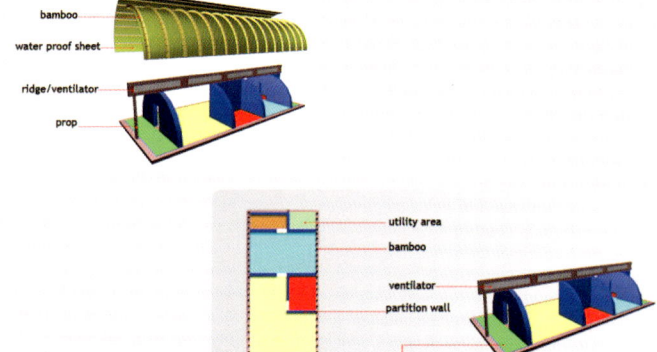

ventilator/window

bended bamboo vegetation

bended bamboo vegetation with leaves

DO IT YOURSELF

AUTO- & BIOGENERATED

MATERIALS & CONSTITUTION

RECYCLED & READAPTED

MODULATED & PRE-ARRANGED

DIGITAL FAB

LEE, Junshick

Cornell University

pipoya3029@hotmail.com

The area of research is a condensed one, about 30% of it is open space, most of it is used as dumping area and it is hard to be kept organized. In contrast to the public areas, moving on to the inner streets, the open spaces are more organized. The typical houses in Mukuru are of one-room, so the inhabitants use semi-private spaces as supportive ones for their other everyday activities. If the sewage line was placed inbetween two houses, almost the 90% of that area would become informal toilet, washing area, laundry or bathroom. Other functions of semi-private space were cooking and storage. By suggesting this semi-private space, we wanted the inhabitants to feel the social responsibility to control this area under the community based participation. In other words, through proposing the semi-private space, we prevented the open spaces from becoming a dumping area. By community based participation, construction is expanded by land and building structure ownership. In each phase, only 5-8 buildings are temporarily re-located.

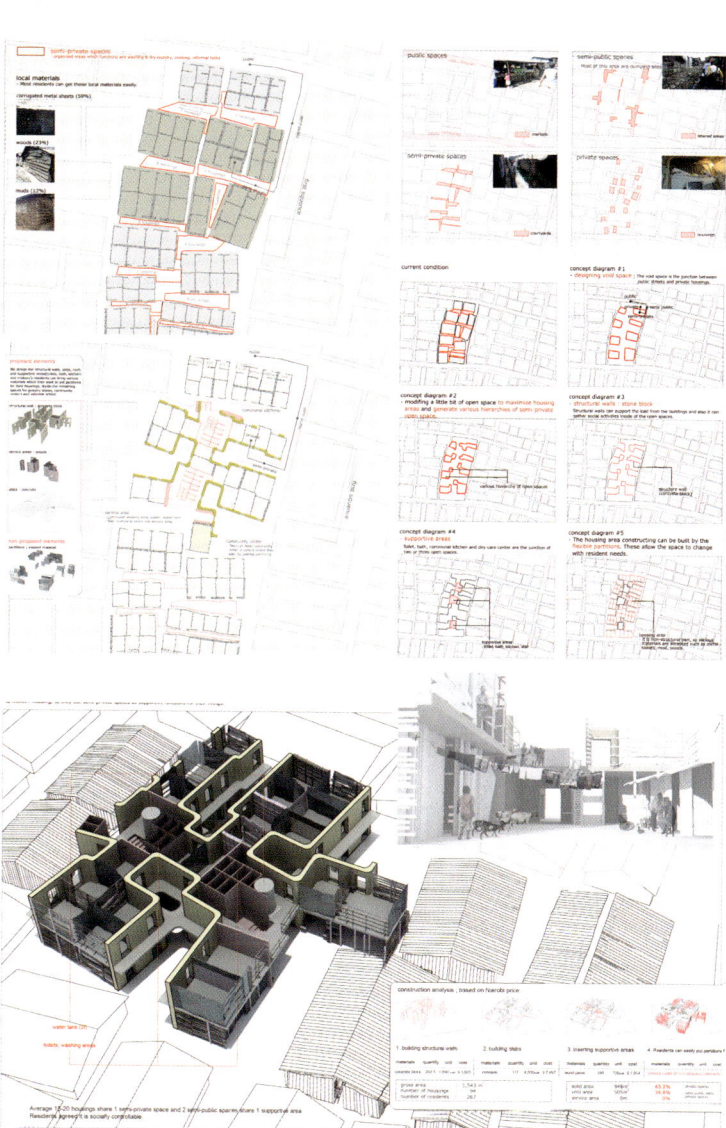

DO IT YOURSELF

AUTO- & BIOGENERATED

MATERIALS & CONSTITUTION

RECYCLED & READAPTED

MODULATED & PRE-ARRANGED

DIGITAL FAB

local materials

corrugated metal sheets (59%)

woods (23%)

mud (12%)

public spaces

semi-public spaces

semi-private spaces

private spaces

current condition

concept diagram #1 — designing void space

concept diagram #2 — modifying a little bit of open space to maximize housing area and generate various hierarchies of semi-private open space.

concept diagram #3 — structure walls - stone block

concept diagram #4 — supportive area

concept diagram #5 — The housing area constructing can be built by the flexible partitions. These allow the space to change with resident needs.

construction analysis : based on Nairobi price

1. building structure walls 2. building slabs 3. inserting supportive areas 4. Residents can easily put partitions !

GARCIA MARTINEZ, Angel   Universidad de Alicante

                                    arq_peirl@hotmail.com

Bamboo Structure:

The structure is made by miscellaneous steel-bamboo, with strong bamboo and steel connections. External skin as plastic fuselage.

Water, Insulation Material:

Constant circuit of water to the interior, thanks to a skin made with "water bags". The building gathers the water from the river, and is re-circulating it in the interior. During the Winter a closed circuit is created and the water is warmed by few electrical resistances, while in the summer an open one protects the interior from the exterior heat. Moreover, during the warmest days the building creates an artificial cloud of pulverized water, making the high temperatures minor.

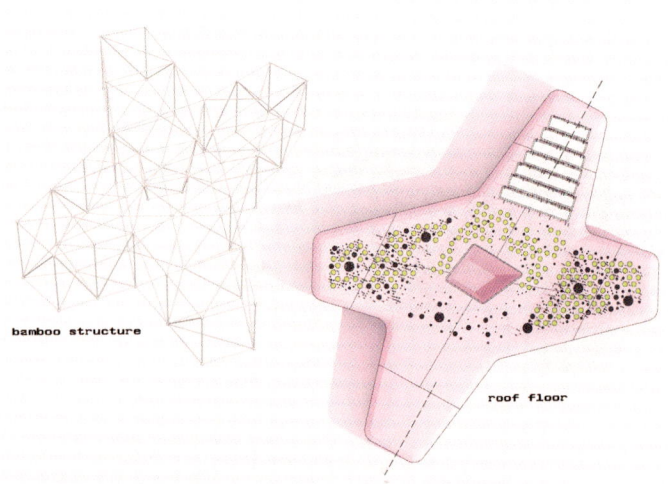

bamboo structure

roof floor

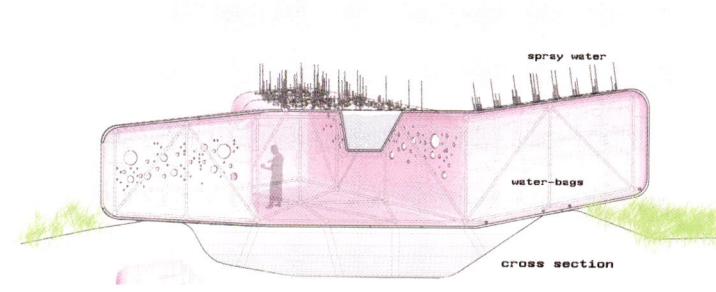

spray water

water-bags

cross section

DO IT YOURSELF

AUTO- & BIOGENERATED

MATERIALS & CONSTITUTION

RECYCLED & READAPTED

MODULATED & PRE-ARRANGED

DIGITAL FAB

BERASATEGUI CANALS, Marina
DIRKMAAT, Remko Jan
STOLK, Anneke

Escola Tècnica Superior d'Arquitectura
de Barcelona

marina.berasategui@gmail.com

The main aim is to divert from the actual direction of things and propose a strategy that shows a new path with a collective contribution.

Tourism is a worldwide massive phenomenon that must self-manage its own waste and be used as a medium for the strategy to be spread.

Travellers consume culture, gastronomy and landscape. Banners, empty bottles and plastic bags can be reused by the same first consumers as a temporary living space.

To elongate the useful lifetime of a product, it is essential to enhance the recycling process without using extra energy for it.

As Barcelona offers tourism items to be consumed, it can also offer its own waste ground floor space as a recycled location.

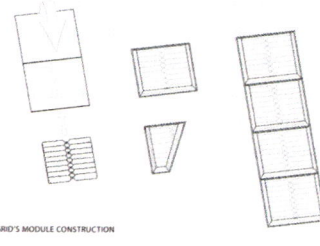

GRID'S MODULE CONSTRUCTION

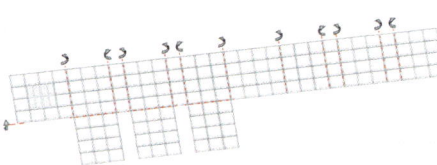

GRID'S DEVELOPMENT AND FOLDING PROCESS

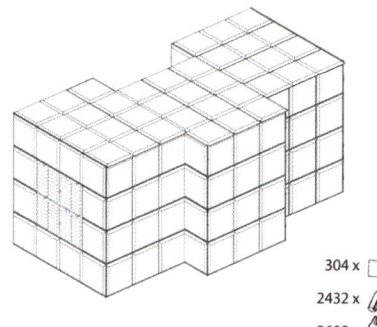

RECYCLED WASTE ITEMS USED FOR A
MODULAR HOUSE

304 x
2432 x
3600 x
6336 x

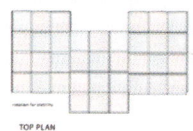

TOP PLAN

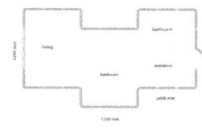

GROUNDFLOOR PLAN

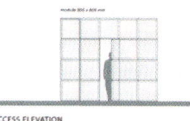

ACCESS ELEVATION

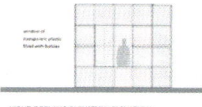

LIGHT OPENING ELEVATION ELEVATION

SECTION

DO IT YOURSELF

AUTO- & BIOGENERATED

MATERIALS & CONSTITUTION

RECYCLED & READAPTED

MODULATED & PRE-ARRANGED

DIGITAL FAB

KUAN YU, Lin

National United University

vrohya@hotmail.com

Neihaizau is actually the Cigu Lagoon which is called like this by the Cigu Fishing Men. The current Lagoon landscape was formed due to the change of the routes of the Zhengwen stream for four times. So far it is the biggest Lagoon in Taiwan. This place has abundant ecological resources.

Rain water will be stored in the tank for use, and the weight of rain water will help the oyster container to rise up easily and protect the stream from pollution. Sea water can be transformed to pure water for daily use and it could be used to fill in living unit shelters. Oysters will be protected by "colloidal space" in order to maintain ocean's bio-diversity. The living units use "colloidal skin" to slack external environment's impact, it also enhance the flexibility of living space. Sea water will be bumped into skin pipe with tide power. This shelter ease heat impact from sun by boiled sea water in skin pipe. This system could recycle boiled water to the tank for use.

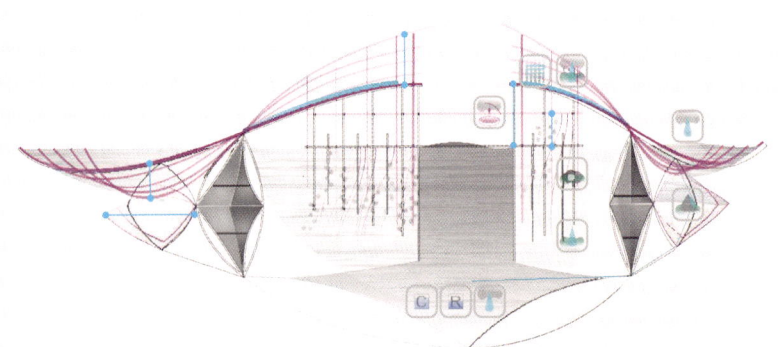

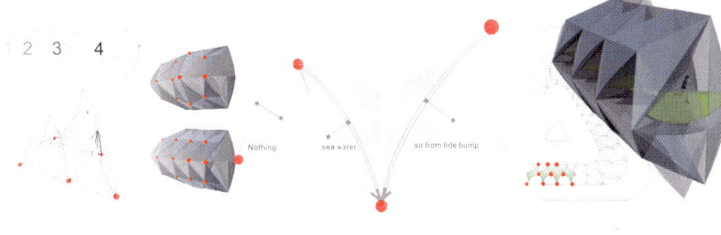

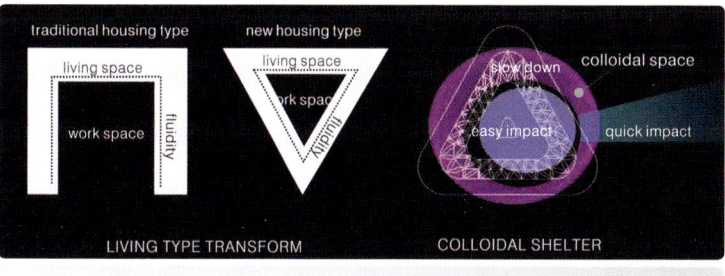

traditional housing type

living space

work space

fluidity

new housing type

living space

work space

fluidity

colloidal space

slow down

easy impact

quick impact

LIVING TYPE TRANSFORM

COLLOIDAL SHELTER

DO IT YOURSELF

AUTO- & BIOGENERATED

MATERIALS & CONSTITUTION

RECYCLED & READAPTED

MODULATED & PRE-ARRANGED

DIGITAL FAB

OZTURKOGLU, Meryem

IMM ARCHITECTURE

meryemozturkoglu@hotmail.com

Darfur is situated in Africa at the Cad border of Sudan. The negative present situation of local people living there is because of forcement of ethnic groups. They all live in refugee camps. Thousands of them are forced to leave their villages. Against to all these negative situations they are trying to live in refugee camps. While some of these poor people are trying to accomodate in tents, some of the others are trying to live in the sheds which they made from pieces of boxes that get from helping organizations. In every aspect it is very hard to live under these circumstances and hardness of the desert as well. In my project i designed "sprining up houses made from adobe brick" in order to over come to accomodation problem.

Conception Criteria

1. Adobe Houses: The main reason for an adobe house to be chosen to be built is its easy construction. Abobe brick making is a simple technology: all that really requires is dirt, water, and a hole in the ground to mix. If you want to understand the consistency of the mix you can make a small ball in your hands and throw down the ball when it doesn't fall to pieces the consistency of the mix is enough to make adobe bricks. In this project 9x26x12cm adobe bricks are used. The mix is left at least one day to be ready for the bricks. After that the mix is poured in to the wooden tablets. When the bricks are starting to be solid the wooden tablets are taken over the bricks. After that the bricks are left under sunlight to be ready for making the house.In case of making the foundation of the adobe house we can use some stones and cement. About 45 cm foundation will be enough for the house.

2. Grammer Based Springing up: Human beings need sociality with others in daily life this sociality has its own hierarchy, we tried to construct it with this grammer based design. In that way the concepts such us neighbourhood houses, social and canopy places could be able to be established. We tried to create a village which is dreamt by Darfur people by using this method.

3. Protecting from the Sun: It is very difficult for a human being to live in the desert due to the harsh climate conditions (strict sun shine in the morning and cold at night). The neigboring of the houses provides a preservation from direct sunlight for their residents. The isolation of the houses is tried to be provided by the frequent branches which are used for roof, from the sun as well.

adobe walls

B
A
B
A
B
A

adobe wall lines

plan of walls

wall facade

ADOBE HOUSES

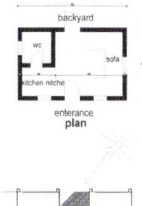

plan

section- being protecred from sunlight

facade

design principles

site hierarchy

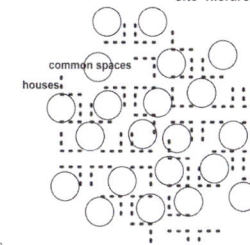

common spaces

houses

grammar based design

adobe house-basic modul

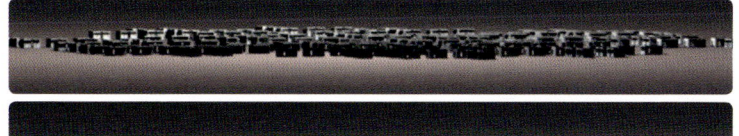

DO IT YOURSELF

AUTO- & BIOGENERATED

MATERIALS & CONSTITUTION

RECYCLED & READAPTED

MODULATED & PRE-ARRANGED

DIGITAL FAB

CALDWELL, Kyle

University of Oregon

manzanitaksc@yahoo.com

Concepts of Self Sufficiency must embrace an integrated approach of social and environmental concerns. True sustainability will not come from a solely introverted expeditions, but will arise from social cohesion and unity. Architecturally, we must strive to engage humanity and the environment together as one. The Muscle + Bones Flexible House embraces the concept of an interconnected economy that excels with integration. The Flexible House exists as a relative discussion-point in the urbanized landscape. Situated in the urban context, Muscle + Bones is allowed to simultaneously respond both to climatic changes as well as daily shifts in social interaction, mediating the degree of privacy and public involvement.

The Muscle & Bones Panel integrates simple tension based mechanical technology with pattern-based overlapping and layering. The system strives to be as integrated as possible, allowing a cohesion of environmental, social, structural, and automated aspects. The automatic fabrication sequence gives way to automated site and user specific adaptation habits. While the tension system continues to move the panels, they also remain to give structural support.

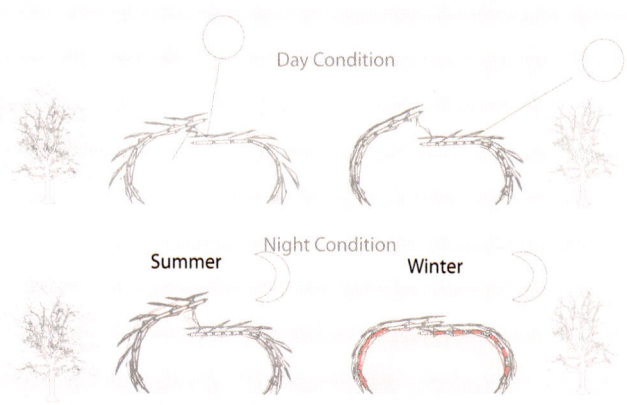

Day Condition

Summer   Night Condition   Winter

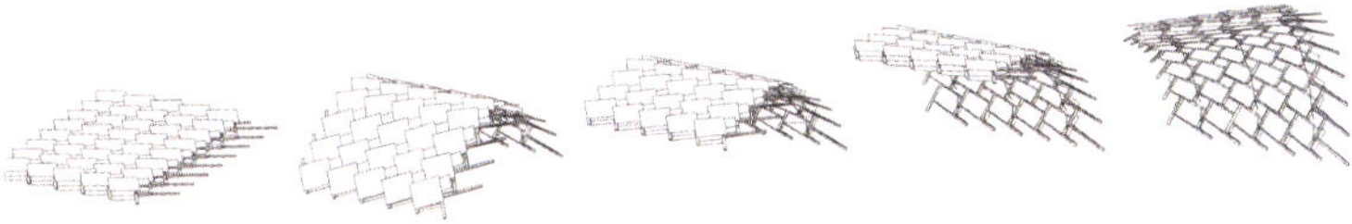

DO IT YOURSELF

AUTO- & BIOGENERATED

MATERIALS & CONSTITUTION

RECYCLED & READAPTED

MODULATED & PRE-ARRANGED

DIGITAL FAB

ARBAB, Kaveh   The Ohio State University

kaveharbab@gmail.com

Since 1990 Vietnam has lost 75% of its forests due to post war industrialization and waste left from the war. The effects have had serious repercussions on both the ecology and communities of Vietnam. The research and conceptual development presented in this project aim to fuse traditional, scientific and architectural ideas in order to confront these issues.

What brings these fields together is the application of carbon aerogels. Carbon aerogel has the capacity to perform multiple functions that are utilized in the proposed system. The system is represented in the community diagram. Energy is collected and produced through a silicon system proposed by Freeman Dyson. This energy is used to fluidize carbonaceous substances, including biomaterials, and produce the aerogel at various micropore sizes. The variation in micropore sizes allows the carbon aerogel to perform its multiple functions. These include filtration, solar energy absorption, and electrical conductivity for functions within the dwelling units. The liquified aerogel is applied to a bamboo frame which is woven together by a mesh produced with traditional vietnamese techniques for strength and efficiency. This allows the aerogel to seep out on to the mesh and form a durable and breathable surface. The varying states of carbon aerogel are sent to specific cell locations within the frames of the dwelling units and may be reapplied there repeatedly when in need of repair.

This system aims to improve the quality of the Vietnames ecology and allow the people to control their means of production. The challenge is to develop this technology and make it available to those who will find it most useful.

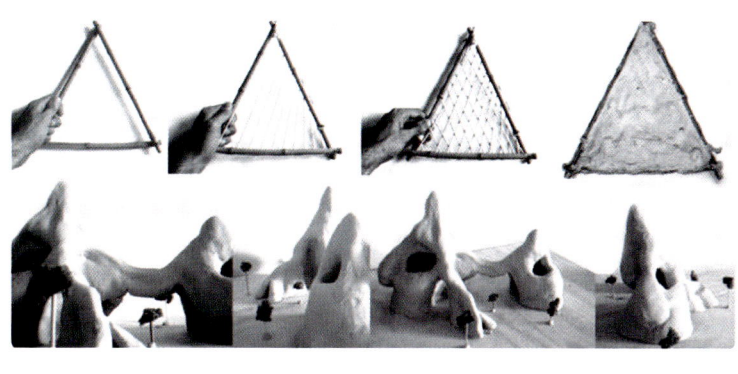

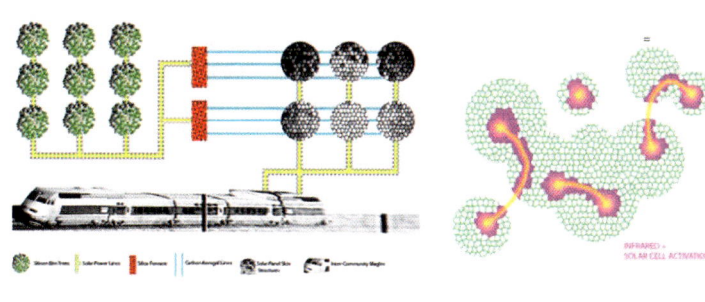

INFRARED +
SOLAR CELL ACTIVATION

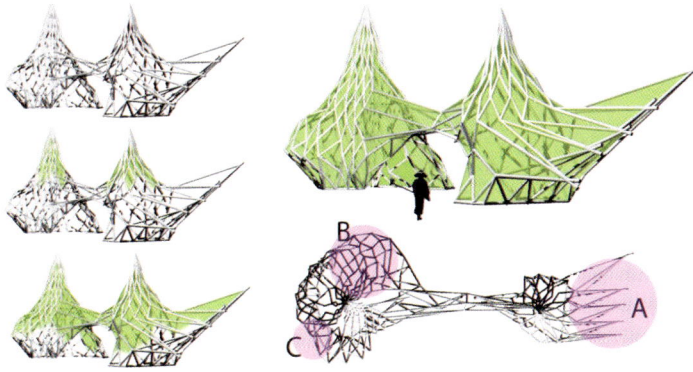

DO IT YOURSELF

AUTO- & BIOGENERATED

MATERIALS & CONSTITUTION

RECYCLED & READAPTED

MODULATED & PRE-ARRANGED

DIGITAL FAB

MOORE, Ashley   The Work.Group

ashley@sisyphean.com

Netbrix is a system of data-sharing construction modules. As an individual, each Netbrix contains the capability to learn its function, location, and orientation with regard to all other Netbrix in a structure. Communicating with built-in WiFi adapters, the Netbrix - which may be transparent, opaque, photo voltaic – are able to record and share information such as energy efficiency, global positioning, vertical and horizontal organization. As a structure is built, the network between the Netbrix builds the corresponding digital map of the structure. Thus, the construction documents are drawn in the real time it takes to put together the house or building.

These construction documents are instantaneously shared across the planet via peer-to-peer networks. Any user wishing to build with Netbrix will have access to a digital repository of every Netbrix structure that has been built up to that very minute. The designs of Netbrix buildings are released under a Creative Commons license; any user may take a design and duplicate, alter, merge, hybridize it as he or she sees fit – and their new design is subsequently released.

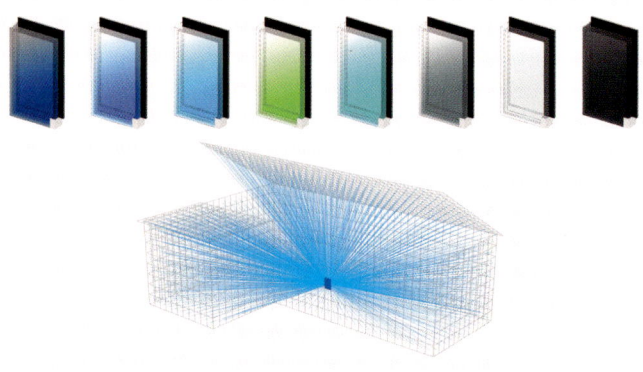

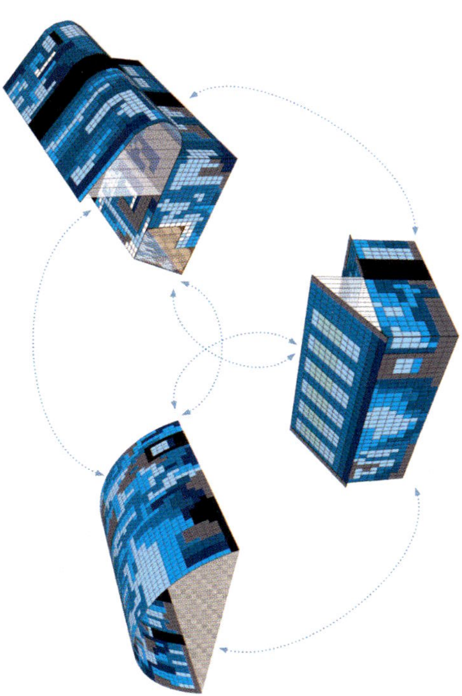

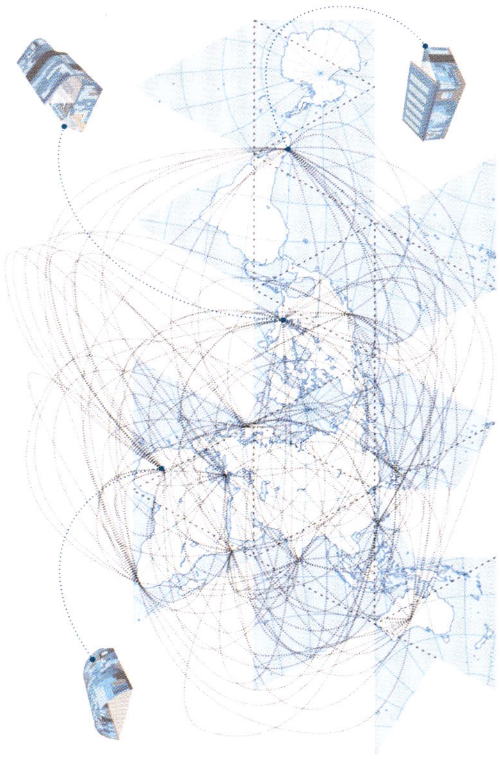

DO IT YOURSELF

AUTO- & BIOGENERATED

MATERIALS & CONSTITUTION

RECYCLED & READAPTED

MODULATED & PRE-ARRANGED

DIGITAL FAB

SPEELMAN FOKKO, Alexander   Hogeschool van Amsterdam, Studio BOT

xanderspeelman@gmail.com

To connect the building to its environment we are interested in investigating the forces that shape this environment. These forces are visible structural, functional, (physical) as well as invisible (cultural, temporal).

Analysis and case studies reveal that the robust design of natural living systems is not produced by optimization and standardization, but by redundancy and differentiation. By adapting to the structural organization of the host, the relation between building and environment will become closer and, ultimately, interactive.

By organizing the building structure through the surrounding forces, such as the river and the mountain's, we try to connect the building to the site. It is an important condition that the border of the building is not a harsh line, but more a connecting intermediary zone that links the building to its host. We will try to achieve this gradient zone by introducing redundancy of structure.

As the environment is continually changing, so should the building. The building's capacity will also change in the future. It is for this reason that we need to develop a structure that can adapt to these changes.

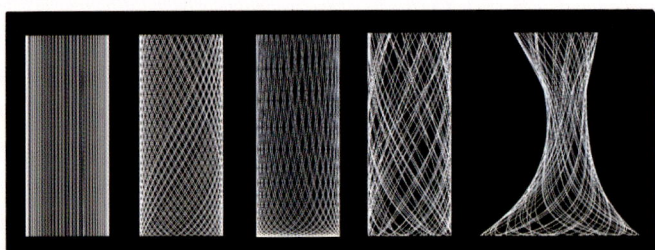

DO IT YOURSELF

AUTO- & BIOGENERATED

MATERIALS & CONSTITUTION

RECYCLED & READAPTED

MODULATED & PRE-ARRANGED

DIGITAL FAB

MAYCON, Oliveira Altera
BRANDAO OLIVEIRA, Tayse Paula

Centro Universitario do Leste de Minas Gerais

maycon.altera@yahoo.com.br

Brazilian rave parties happen in open areas especially those were the contact with nature is possible. It is known to be organized in places near waterfalls, small farms and other areas far from urban centers. it is usually a 12 hour duration event, where DJs and other performers and artists present their music and work, interacting with the public. They dance, jump, cry out and play with glowsticks or juggle.

The "Housing for DJs"structure was made of wood (Eucalyptus). The reasons for taking this material option are because it is reforested wood that has a short life circle and a continuous renewal. The preservation of its physical and mechanical characteristics is guarantied by application of chemical products that retard its decomposition by biological agents.

There have been used 3 meters of eucalypt wood. The canvas cover can be moved to propose an opening depending on the preferences of the user. It has a rigid part where the DJ will be once he is performing and a flexible one where the user can be resting. It is not easy to be disassembled, but once it is assembled offers to the user security, comfort and stability.

DO IT YOURSELF

AUTO- & BIOGENERATED

MATERIALS & CONSTITUTION

RECYCLED & READAPTED

MODULATED & PRE-ARRANGED

DIGITAL FAB

ALCHIE, Laure
DAOUD, Cesar

Ecole Nationale Supérieure d'Architecture
et de Paysage de Bordeaux (ENSAPBx)

laure.alchie@hotmail.fr

When we think about self-sufficiency, we usually refer to the numerous stereotypes connected to the rural society and the natural elements. But when we realize that in 2020, 80% of the population is going to live in the city, we are entitled to wonder what this notion means in the urban environment - our surrounding being only made by the men and for the man.

This project proposes a structure that coincides after the agreement between the authorities of a city, a communication firm (that deals with marketing) and homeless people living at the streets and finally offers shelter to the last ones. The city provides material as long as space and infrastructure, the advertising company offers the money and finally the people are fabricating their house and are living in it. This project tries to combine these totally different elements in order to generate a new living space in the streets of the city. Additionally, it aims to change our collective conciousness about our responsabilities in the city.

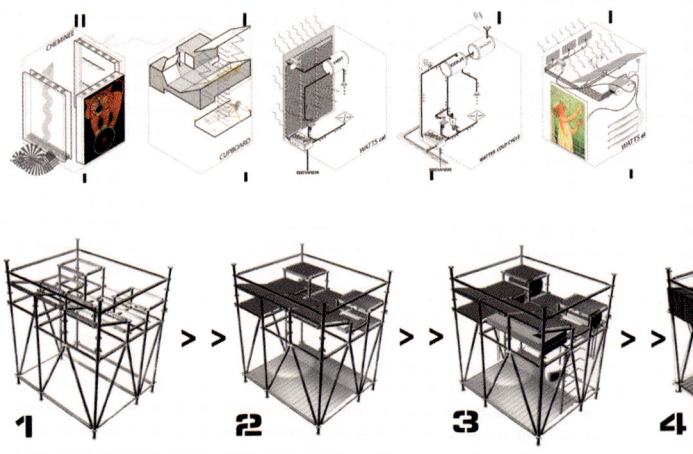

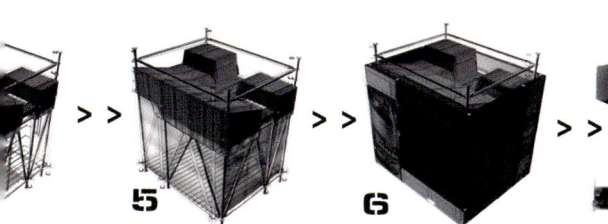

DO IT YOURSELF

AUTO- & BIOGENERATED

MATERIALS & CONSTITUTION

RECYCLED & READAPTED

MODULATED & PRE-ARRANGED

DIGITAL FAB

GRUSS, Hendrik
BEISSERT, Uta

Rafael de la Hoz Arquitectos

hendrik_gruss@gmx.de

All in all, living conditions will decrease to a further minimum of well-being. The struggle for a worthwhile living spot will be harder for every citizen of every age. The city will not be able to offer a place to stay for everyone, instead of persisting the "promised land" cities will run the risk of becoming wastelands.

The scenario: inhabitants - likely pensioners and twens at first - of the mediterranean regions, where the beamiest: effect of global warming will be expected, will start an exodus northwards towards more moderate, livable surroundings - here for instant the so called "polder-landscapes" of the netherlands and northern germany, as participants in our pioneers project - quasi an exodus for ejido - they will find a sparsely populated but subdivided patchwork landscape as their new homeland close to nature.

Finally "exodus for ejido" wants to highlight a more ecology-minded and megapolis-alternative way to live. We want to show up an uncomplicated way to offer space for a complex melting pot of differentiated neighborhoods, in which every grid unit could provide new experience horizons.

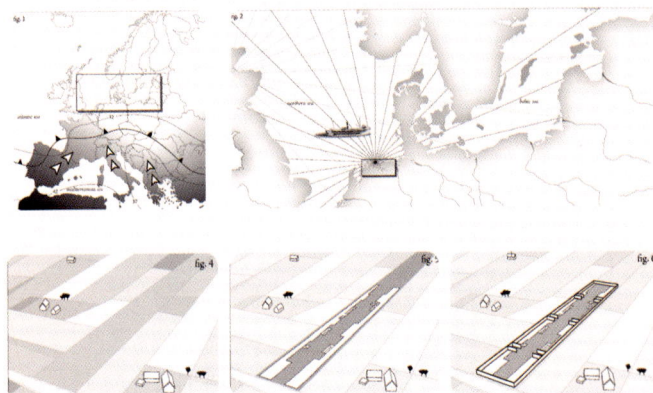

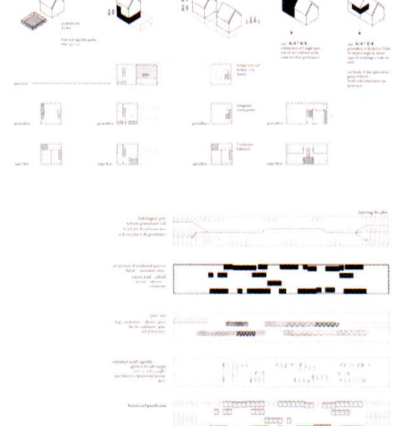

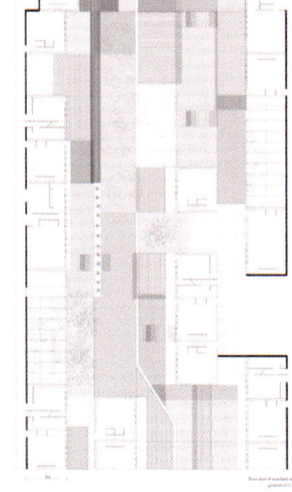

DO IT YOURSELF

AUTO- & BIOGENERATED

MATERIALS & CONSTITUTION

RECYCLED & READAPTED

MODULATED & PRE-ARRANGED

DIGITAL FAB

OKUDA, Shinya
HO ALVIN, Kung Yick
YU IAN, Lam Yan

Department of Architecture,
The Chinese University of Hong Kong

shinyaokuda@cuhk.edu.hk

3RD PRIZE

BVMS_ Biodegradable Vacuum formed Modularized Shelter

Geometrical exploration will start one of the most simple geometry, Vault, with half-circle section. While maintaining same radius would ease of production, the shape is structurally not optimized, so that the vault will take bending moment. The span is 6m. The length is adjustable as needed.

Biodegradable plastic is made of corn, degradable within 20 days into soil under specific condition. The production of Biodegradable plastic has been started in 2004 in PRD, China, currently mainly used for food container with vacuum forming processing.

Vacuum forming is widely used is packaging industry, while some of large scale components can be applied for architectural use, such as Poly carbonate top light dome, etc. The maximum production could be up to 1.2 x 2.4m with 3-5mm thickness. Most of thermoplastics could be applied theoretically, although its rigidity and flexibility are varing depending on types of plastics, such as PS, PET, PVC, etc. The merit of vacuum forming is that once its mould is done, vacuum forming process is relatively easy and it could be self-fabricated.

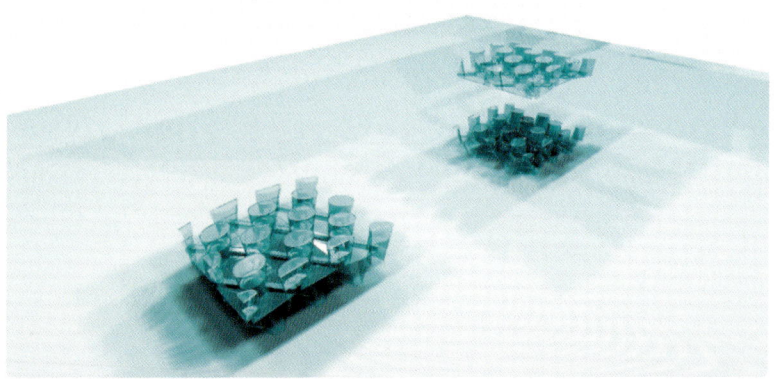

SECTION DETAIL

INNER MODULE SECTION

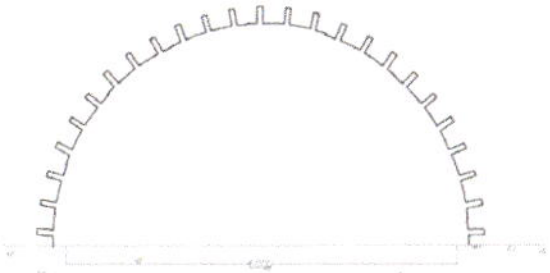

OUTER MODULE SECTION

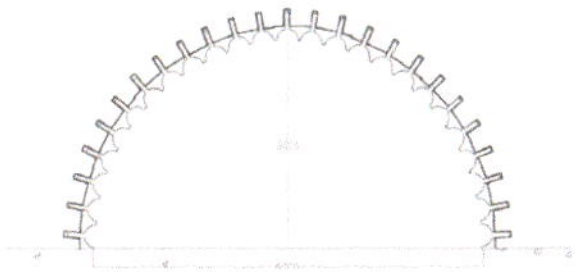

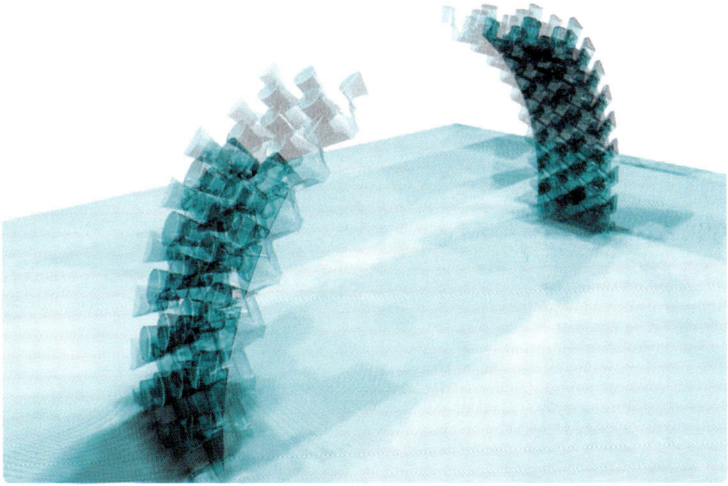

DO IT YOURSELF

AUTO- & BIOGENERATED

MATERIALS & CONSTITUTION

RECYCLED & READAPTED

MODULATED & PRE-ARRANGED

DIGITAL FAB

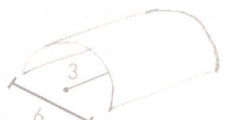

1. Single layer

Corrugated sheet

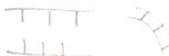

Double layer vault

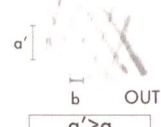

1. Single layer.

$$a' > a$$

2. Double layer

b   OUT

b   IN

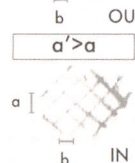

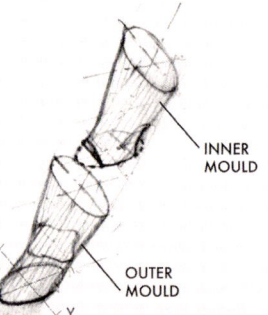

INNER MOULD

OUTER MOULD

Y

2. Double layer

a. Dot pattern   b. Orthogonal grid   c. Lattice grid

Elevation

Plan

Section

DO IT YOURSELF

AUTO- & BIOGENERATED

MATERIALS & CONSTITUTION

RECYCLED & READAPTED

MODULATED & PRE-ARRANGED

DIGITAL FAB

DO IT YOURSELF

AUTO- & BIOGENERATED

MATERIALS & CONSTITUTION

RECYCLED & READAPTED

MODULATED & PRE-ARRANGED

DIGITAL FAB

RECYCLED AND READAPTED

The more the population of Earth is increasing, the bigger the effect on the environment. The recycling of materials, products or even objects is something extremely vital, so as for the waste disposal on earth to be as diminished as possible. This category is exploring the various possibilities of fabrication of a building by its own user, by recycling materials, products or even objects.

Technological Advances are applied in the construction industry; materials with specific properties that make them recyclable are being used in the building structure. The use of recyclable materials as basic materials of a structure is something that could be considered very important, due to the fact that the possibility of their destruction will not leave any waste; the material could easily be recycled a series of times.

Additionally, it is possible that objects or parts of other structures that already have lost their basic functional characteristics could be recycled, reconfigured and re-adapted as important structural parts of a dwelling.

These could result in low-cost structural solutions, due to the fact that there is no need for new materials or objects or structural parts to be produced and no specific need for specialized labour. Finally, the customization and personal expression of the user on the construction of its own building is extremely promoted. It is very interesting the fact that the same object can be proposed as a modular part, resulting in a series of different construction systems.

The use of recycled objects and their re-adaptation in the building construction is something that has been already performed by specific cultures. The consideration of these lifestyles as important, as long as the intention to learn more from them, is a step-forward in to order to built more sustainable systems of housing. The influence by traditional principles and materials in the construction, as long as their combination with the technological advances offered nowadays, can be very effective in the contemporary architecture.

DO IT YOURSELF

AUTO- & BIOGENERATED

MATERIALS & CONSTITUTION

RECYCLED & READAPTED

MODULATED & PRE-ARRANGED

DIGITAL FAB

SURVIVAL KIT

SUNGIL, Kin
JAEIN, Choi
HUN, Kang

Forest Architects

sungil17@yahoo.com

Global warming caused by environmental pollution leads to tremendous damage in many parts of the world. The damage becomes larger and unexpected results are appearing in succession. Warmer temperatures could also increase the probability of flood, more frequent and intensive heat wave. Moreover, warmer water in the oceans pumps more energy into tropical storms, making them more intense and potencially more destructive. Our innovative shelter is designed for victims of disasters who lost their homes and their livelihoods by a sudden disaster. The shelter is earth-friendly, non-toxic and can be assembled easily and quickly in disaster area. The surface material of the shelter is made of 25% of recycled paper dressed with high density Polyethylene that has been featured with water proofing, vapor transmission and radial heat protection. This environmentally friendly material is 100% recyclable and it only discharges water vapor and carbon dioxide when destruction by fire.

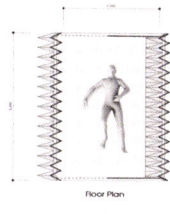

Floor Plan   elevation   Section

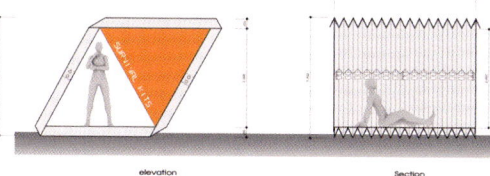

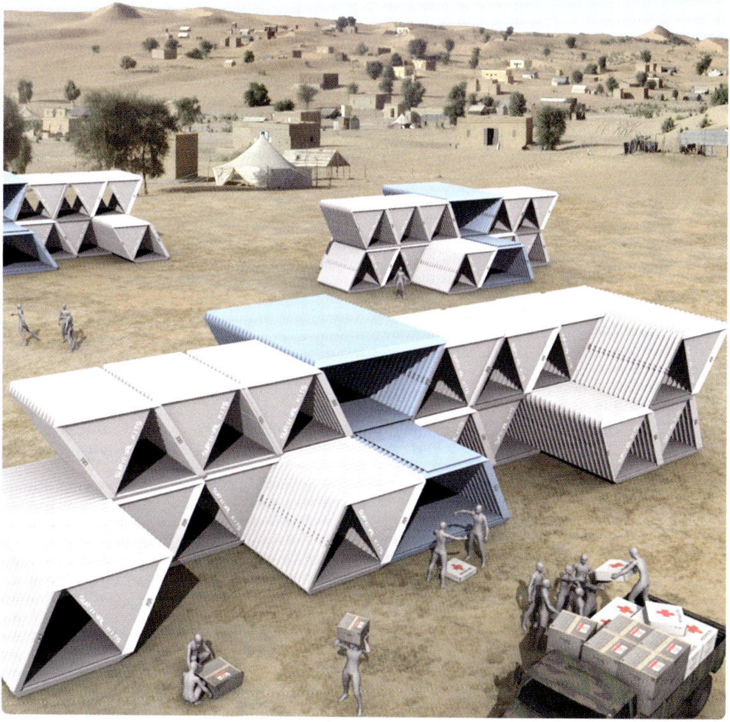

DO IT YOURSELF

AUTO- & BIOGENERATED

MATERIALS & CONSTITUTION

RECYCLED & READAPTED

MODULATED & PRE-ARRANGED

DIGITAL FAB

PROPOSAL FOR WATER IRIS PROBLEM

Mexico

DOMENZAIN, Carlos
SANCHEZ ROJAS, Gerardo David

ITESM, Campus Guadalajara

cadomenzain@hotmail.com

The relationship between the local materials and the concept of a self-sufficient lake community are the leading elements that configure this project.

Modules are based on a systematization of wooden boxes used as temporary containers in regional markets. The module acts as a reference of a contained space that can be modified, transformed, depending on the subjects living inside it and on the activities that can be developed by them. Consequently, the module acts like a "cell", a living organism that is interconnected with others: the community idea. This community is related through a self-sufficient development of the basic needs that are based on two principles: water regeneration and energy obtained from organic wastes.

Here more problems are solved: the overpopulation of water-iris that can be used as key element to obtain purified water and also as an organic fuel.

In order to link the modules and create a "multicelular" organism a common system of purified water and energy can be developed. This systems, that exist as an individual entity, would proliferate as an alternative way of life that could act, along with the lake, in a symbiotic reltionship, solving, in the future, iris overpopulation and pollution problems.

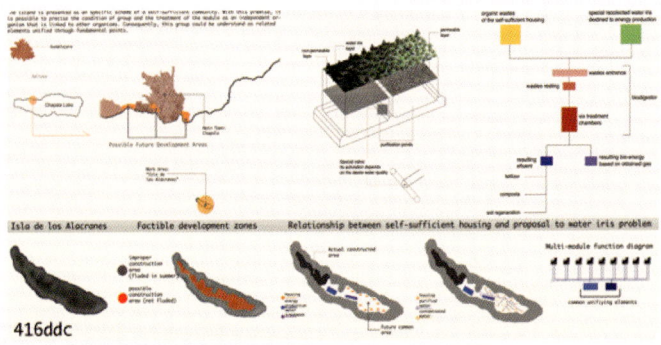

Isla de los Alacranes Factible development zones Relationship between self-sufficient housing and proposal to water iris problem Multi-module function diagram

416ddc

DO IT YOURSELF

AUTO- & BIOGENERATED

MATERIALS & CONSTITUTION

RECYCLED & READAPTED

MODULATED & PRE-ARRANGED

DIGITAL FAB

2CESD1

TORRE SANCHEZ, Roman
KUKUCSKA, Gergo

Self-taught designer

vidainutil@gmail.com

What is this? It is an offer of a basic module made of recycled material, inflatable and easy to transport, disposing of a whole series of mini-facilities that allow self-sufficiency, both energetic and personal.

This is a project in which, obviously, we have tried to confront our different social and vital realities (Budapest, Hungary vs Gijon, Spain) as well as other realities unknown to us, approaching the subject of the form more universally than has so far been for us, of course opening towards and focusing on actual and hypothetical local situations.

Where and how? You can install the modules off different ways: on your balcony or terrace, or you can send it to other places, for example as on emergency solution for countries of latin America, or in case of natural disasters. The module could be a convenient choice for a North Pole expedition, in which case may also need a useable climatic jacket, but we are sure that the self-inflatable house could prove helpful to confront the severe conditions in those distant and cold places.

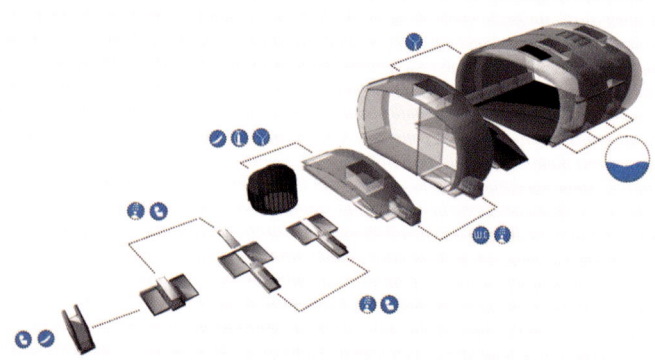

DO IT YOURSELF

AUTO- & BIODEGENERATED

MATERIALS & CONSTITUTION

RECYCLED & READAPTED

MODULATED & PRE-ARRANGED

DIGITAL FAB

A1

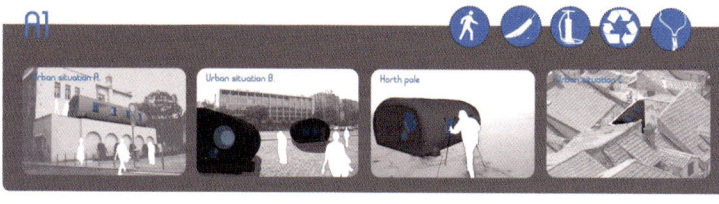

Urban situation A. Urban situation B. Harth pole Urban situation

A2

Same views   Inside view   Simbols   Inside view

Simbols
Assembly — Shower
Inflate — Transport
Water tank — Zipper
EcoUTC
Residual tank

A3

Variations   Simbols

Simbols
Inflate — Zipper
Rain water tank — Recycle
Water tank
Solar Power
Residual tank

VALIENTE ORIOL, Gonzalo   Escuela técnica de Arquitectura de Madrid
VALIENTE ORIOL, Jorge
                                    pancrudino@hotmail.com

One of the main ecological problems nowadays, as a consequence of the non-stop activity in everyday life, is the accumulation of waste products. Many of the waste products are underestimated by the society, they can be reconsidered and used with a new sense, so as to finally be redefined. Our intention is to propose a new activity, redefining a waste product. Our investigation starts by considering the large number of cars and vans that are used every day. Their "life" period will sortly end and they will leave a huge ammount of waste. The recycling of the metallic components, although can be done, is energetically expensive.

The shelter is built by the components of 3 vans and 1 car. We selected the IVECO vans because they are the ones mostly used in the market and they should have a close future. The car used was a fiat 126 p. The top seats of the car are redefined as stairs. And the other half of it is used as a seat for 2 people, not only it can be used as a dining area but also it extends the internal space of the shelter.

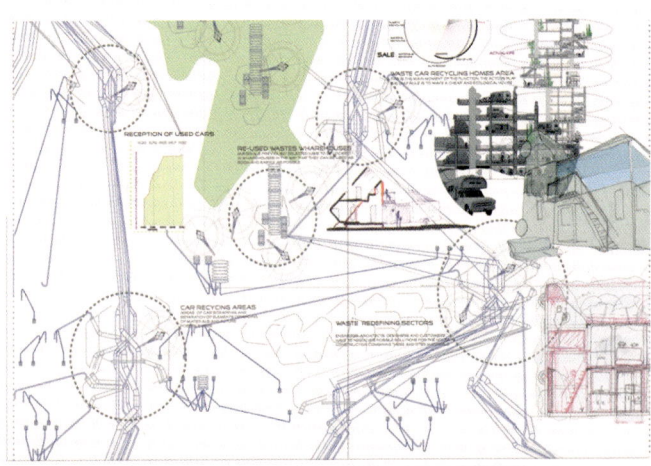

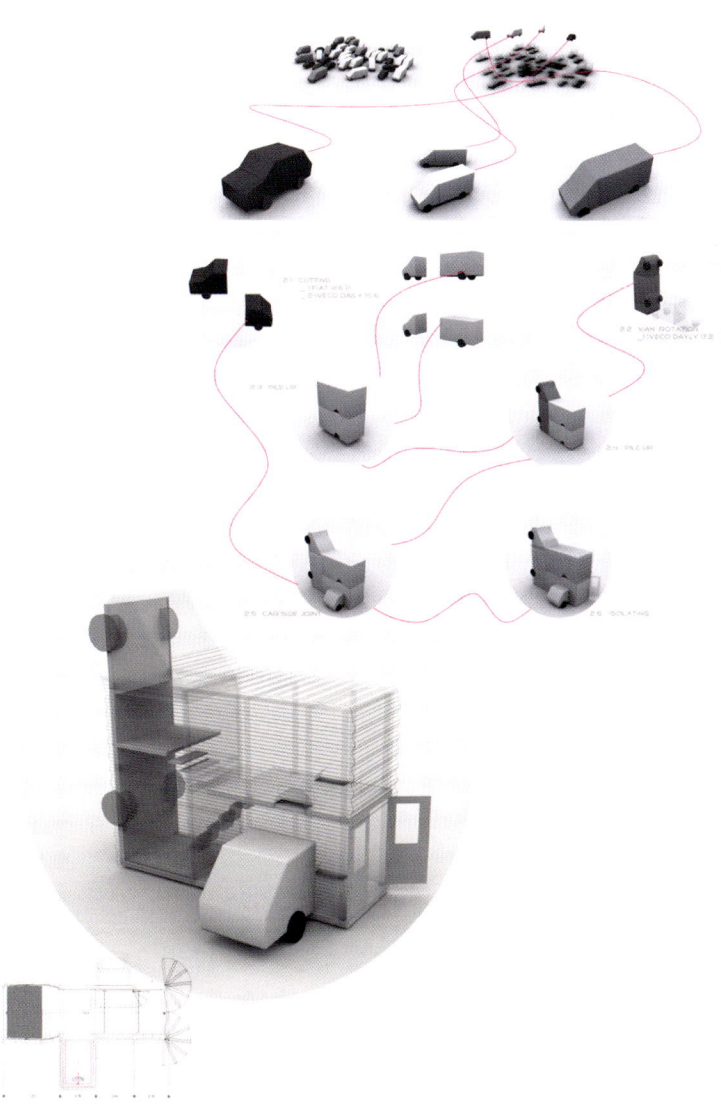

DO IT YOURSELF

AUTO- & BIOGENERATED

MATERIALS & CONSTITUTION

RECYCLED & READAPTED

MODULATED & PRE-ARRANGED

DIGITAL FAB

VECINO, Victor

ETSArquitectura A Coruña

victorvecino@hotmail.com

The need_ thousands of people wander around cities without any means, without a roof over their heads. The starting point of the research of this proposal is a shelter.

Recycling_ our landscapes are spotted with huge used-tyres, yards waiting for somebody to find something to do with them. Shall we re-use them?

The wheel_HOUSE is made up of 9 tyres of the same size. When you cut them in halves you get 18 pieces. These 18 pieces are joined with passing screws adjusted with nuts which make the structure rigid. At the same time, this system allows you to move the structure in order to get the positions of transport and use.

When the shelter is disassembled it is converted into a 1.30 m diameter wheel, inside it luggages can be placed. It is needed only 30 seconds for the bed to be disassembled. Both sides can be covered with a plastic re-cycled material, aswell, so as the user to create privacy and be protected from the wind and the rain.

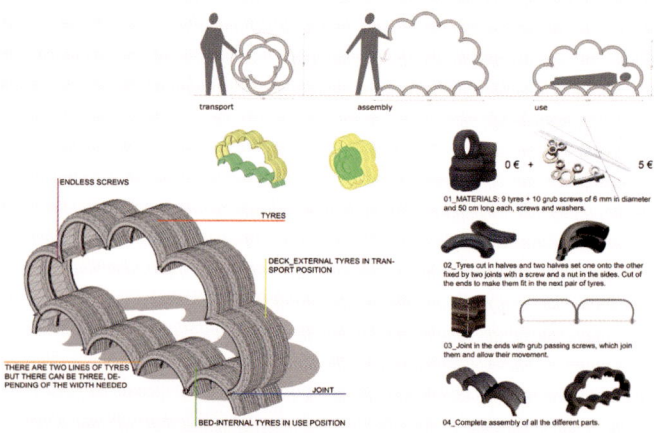

transport   assembly   use

ENDLESS SCREWS

TYRES

DECK_EXTERNAL TYRES IN TRAN-
SPORT POSITION

THERE ARE TWO LINES OF TYRES
BUT THERE CAN BE THREE, DE-
PENDING OF THE WIDTH NEEDED

JOINT

BED-INTERNAL TYRES IN USE POSITION

01_MATERIALS: 9 tyres + 10 grub screws of 6 mm in diameter
and 50 cm long each, screws and washers.

02_Tyres cut in halves and two halves set one onto the other
fixed by two joints with a screw and a nut in the sides. Cut of
the ends to make them fit in the next pair of tyres.

03_Joint in the ends with grub passing screws, which join
them and allow their movement.

04_Complete assembly of all the different parts.

9.00am

9.30am

11.00am

12.00pm

13.30pm

15.00pm

20.00pm

08.00am

DO IT YOURSELF

AUTO- & BIOGENERATED

MATERIALS & CONSTITUTION

RECYCLED & READAPTED

MODULATED & PRE-ARRANGED

DIGITAL FAB

AGUIRRE MANSO, Luis

AQSO

info@aqso.net

2ND PRIZE

The self-sufficient housing concept is especially relevant in the areas where there is a strong demand for qualified labour and technical resources are scare.

In the last 40 years, approximately 3 million people have been forced to leave their homes in Colombia, quantifying it as the country with the highest number of displaced people in the world.

The project is located in the Cauca valley, between the Occidental and Central mountain chain, where a superb construction material is abundant whose growth is natural: bamboo. This versatile, light-weight, biodegradable and sustainable raw material differs from other woods in the fact that it grows again after being cut.

The "Harvest Home" is an experimental house based on a traditional scheme: day and night activities are separated in two floors around a fireplace. Its slender figure becomes an organic element that rises over the landscape's valley and its bamboo shell creates a hard wearing structure interwoven with a wicker mesh.

Houses in Cauca valley are ready to be harvested. Their occupants will have a hand made and sustainable space to live in, while the bamboo will continually replenish itself in this fertile Colombian region.

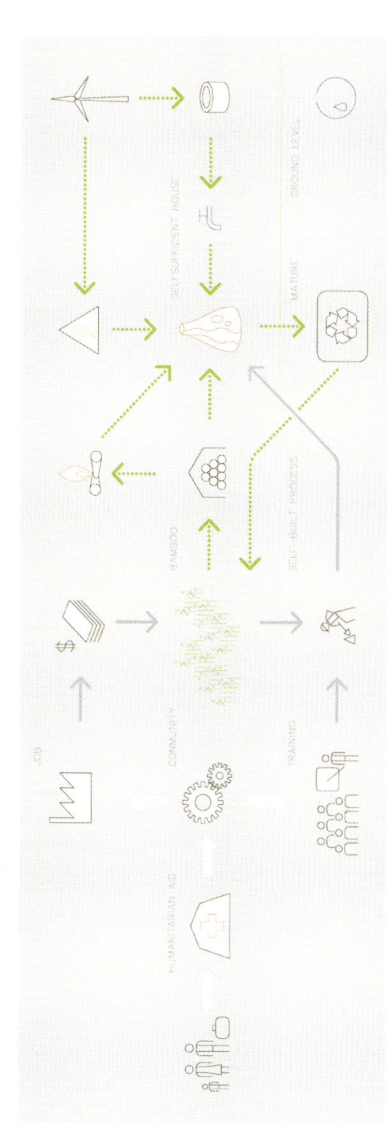

DO IT YOURSELF

AUTO- & BIOGENERATED

MATERIALS & CONSTITUTION

RECYCLED & READAPTED

MODULATED & PRE-ARRANGED

DIGITAL FAB

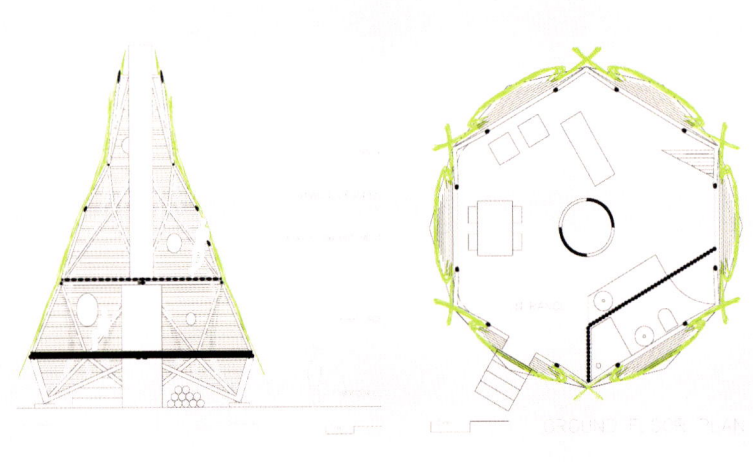

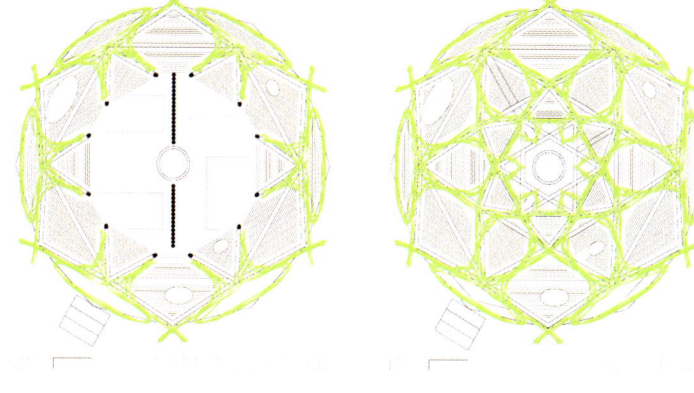

DO IT YOURSELF

AUTO - & BIOGENERATED

MATERIALS & CONSTITUTION

RECYCLED & READAPTED

MODULATED & PRE-ARRANGED

DIGITAL FAB

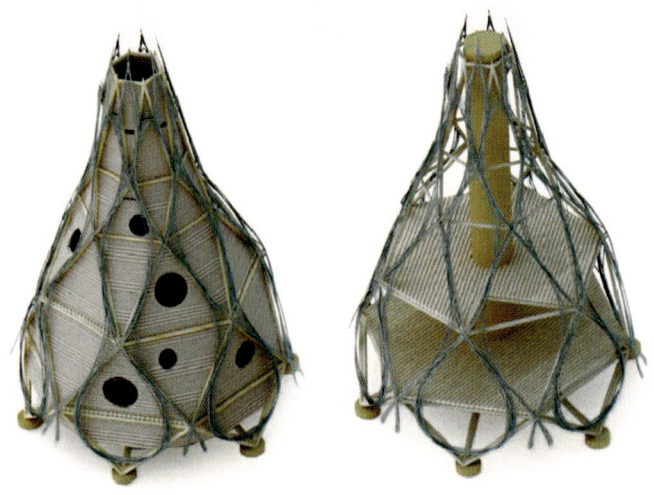

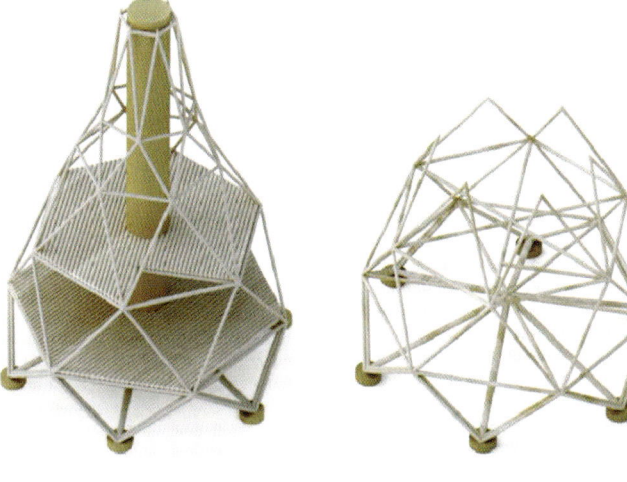

DO IT YOURSELF

AUTO- & BIOGENERATED

MATERIALS & CONSTITUTION

RECYCLED & READAPTED

MODULATED & PRE-ARRANGED

DIGITAL-FAB

CABANAS, Miriam   ETSAV
FARNÓS, Gerard Bertomeu
                                     miriamcabanas@gmail.com

CURRENT SCENARIO

Spain is going to exceed kyoto assignments. The estimation of spanish global emissions in 2007 is 400.70 millions of CO_2 tonnes. Spanish ecological organizations defend high-density suburbs but, in fact, spanish coast is a continous sprawl city. This kind of permanent tourism is one of the most important economic factors for the spanish economic growth.

SYMBIOSIS

The project becomes a self-sufficient mechanism for a microproduction of wellbeing and economic benefits. The net of interests sets the system. We propose an algae productive envelop. Like wood or straw traditional building systems, it becomes a CO_2 drain and contributes to slow down the climate change.

CUSTOMER

The product is thought in order to be competetive within the context of a new phase of the economy, where profitability is counted instead of money in energy. The user is making a profit from the system. He's renting a facade that produce energy. So, it becomes a cheaper removable constructive system. One of the main interests during the developing phase will be to take care on the way the product is going to be inserted in the market as long as on how is going to generate confidence. Our aim is to link new producing economies to the traditional dwelling market.

SYSTEM

The link of the water cells to the structure is achieved through the manual assembly of the four pieces that shape the knot. One of the requirements that have been set in the process of design is that this union is easy to implement, as if the user of housing could self-construct an envelope with this system. The knot incorporates two independent networks of facilities: on one hand the conduct of water that feeds each one of the cells. On the other hand, the fiber optic network provides artificial light to the system, in order to keep active the photosynthesis process. We propose, simultaneously, several algae growing developments.

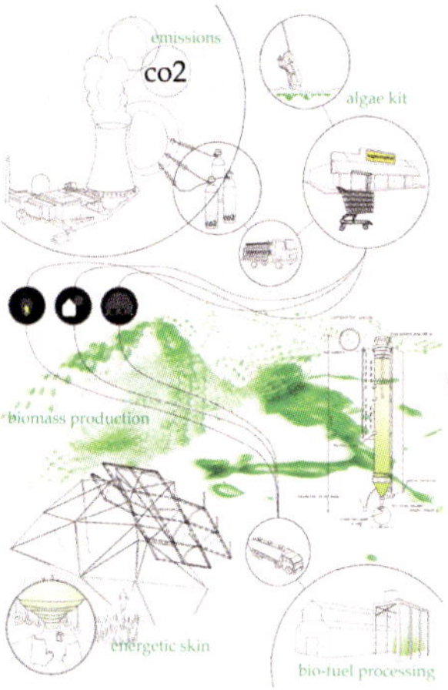

emissions
co2

algae kit

biomass production

energetic skin

bio-fuel processing

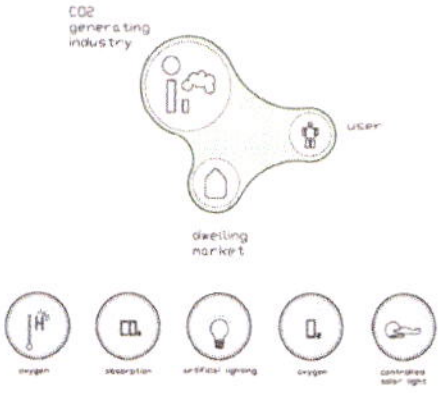

CO2 generating industry

dwelling market

user

DO IT YOURSELF

AUTO- & BIOGENERATED

MATERIALS & CONSTITUTION

RECYCLED & READAPTED

MODULATED & PRE-ARRANGED

DIGITAL FAB

TEIGAN, Annsaint

annarchytect

teigan@sbcglobal.net

Wherever the location, there are elements in abundance. Instead of importing building materials from remote locations, we can manipulate what is already onsite to create shelter. By reproducing low cost moulded components from these plentiful raw materials, everyone can maximize benefit from ready resources. Were we to agree on a worldwide compatible geometry and interface system, even smaller isolated settlements could share in the ever rewarding economy of scale. Engineered components such as form fitted double glazed window panels manufactured en mass, could be made affordable to all, including the least fortunate.

The ideal target structure would be self supporting during construction, eliminating the need for temporary scaffolding and cranes. If a structure is modular in nature, we need only to produce a finite variety of shapes, requiring less tooling. The plethoricity scheme employs only two major tools. These are durable semi-rigid silicone moulds from which bricks can be formed in. The larger mould spans approximately 2 meters and is also the main alignment tool during construction. It takes four workers to extract and place the finished part, but construction ease justifies the large component size. It is pentagonal in shape and a maximum of 6 are used per base structure. In most instances, these will be substituted by lighter pre-fab door, window and ventilation components of the same dimensions. Thus, the larger mould will be used only sparingly. The smaller mould is triangular in shape and is easily operated by a solo worker. A maximum of 54 extracted parts are used per base structure, so multiple of this tool would speed production significantly.

Each brick posesses a hefty wall thickness; both for positioning stability and thermal mass. No permanent mortar is required, although waterproofing is a plus. Absent bricks will not decrease structural integrity provided arches are maintained. Thus, configurations can be easily modified or expanded at will. Upgraded parts will develop through worldwide usage and older parts can be reused for additional expansion. By chaining multiple domes with a small universal arch fixture, we can evolve assymetrical megastructures suited to yet unforseen needs.

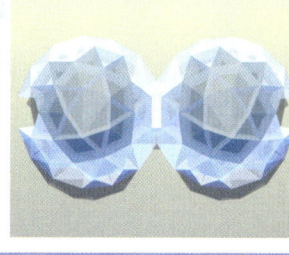

DO IT YOURSELF

AUTO- & BIOGENERATED

MATERIALS & CONSTITUTION

RECYCLED & READAPTED

MODULATED & PRE-ARRANGED

DIGITAL FAB

MUÑOZ BAHAMONDE, Pablo Javier
ZAMBELLI, Jose
CANCINO, Eduardo
LIRA, Ismael
MONZON, Nazca

YZ I AB

pablo_jmb@hotmail.com

In a future where the human actions will be concentrated in more basic the human necessities will result as products after a series of factors that will be affecting the Earth life. In the near future they will be determining the generation of the life of the man and how this is developed in the world.

The return to the essential will be the fundamental thing. For this the self sufficient house appears like a solution before global problems. The selection of a lease, the materials and of a system of suitable operation become essential. A lost island in the south of the world, in Chile, the country with more islands in the world, appears like the ideal place to implant the house. This because the Noir island is remote of the continent, but simultaneously is accessible byboat. The cost of this type of islands is absurd when it is compared it with the value of a department of unifamiliar located in the capital of Chile, since it is much smaller. The objective of the project is that this is not but a problem, we try to expand the globalizacion of the information and the communications, so that the "house-islands" are connected with their immediate surroundings and the entire world.

For the year 2050, when the nano robots technology becomes reality, the special suits will represent a fundamental element for the correct operating of the house and the person itself. These suits provide the optimum conditions for people, in a world that's not longer optimum. Since the suits handle from people's DNA up to information relating states of mind and emotions. This allows the suits to know all the requirements of human beings and how to satisfy them. In addition to that, the special suits are the remote control needed to control in a 100% the house, because it's from the suits that the nano robots rise to fulfill specific tasks in terms of house operations. The suits work thanks to solar energy.

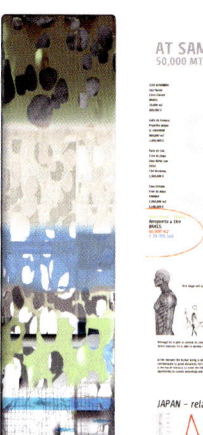

AT SAME PRICE
50,000 MT2 VS 71,5 MZ

ECONOMY

XII REGION
LOW POBLATION

NET COMMUNICATION
DESCENTRALIZATION

JAPAN – relation

LAYERS
SELF FAB HOUSE

COCHAYUYO

MASS
CINR

MASS
BIOREACTORS

MASS
CINR

BIOREACTOR
WATER

HOUSING

TIERRA

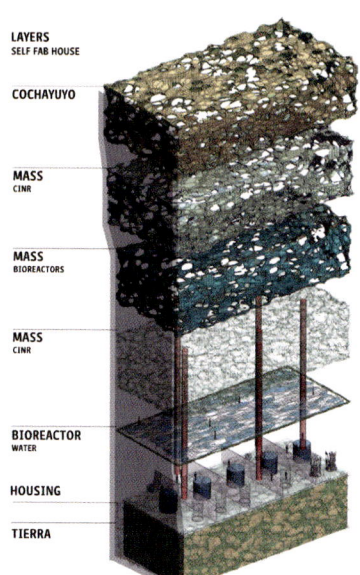

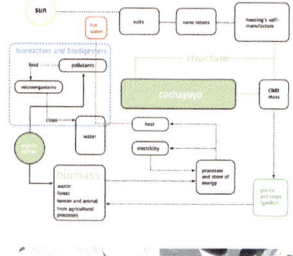

DO IT YOURSELF

AUTO- & BIOGENERATED

MATERIALS & CONSTITUTION

RECYCLED & READAPTED

MODULATED & PRE-ARRANGED

DIGITAL FAB

CAO, Yuan
YAN, Li

Xi`an Architecture& Technology University

caoyuan_531@yahoo.com.cn

Bamboo is a kind of conventional plant in the south of China and is used to make bamboo mat as a living good in this area.

Bamboo is cut into very thin and slender strips, which are then weaved and create a piece of mat. Bamboo mat has great physical strength, durability and toughness.

We braid a long bamboo mat and fixed some iron sticks on the mat along the short side homogeneously.

The tip of each iron stick is shaped into a loop which can make up a knot with a rope. As a result we can use the tension of bamboo mat and the pull of the rope to make the bamboo mat into different shapes.

The new bamboo mat can be used as a temporary pavilion. We can play in it, have a rest in it, and also the new bamboo mat can be easily build up and stored ,so we can even live in it in some emergency situations.

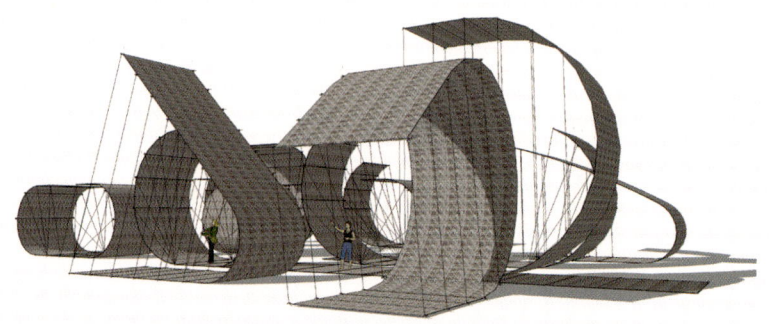

inertia-1

DO IT YOURSELF

AUTO- & BIOGENERATED

MATERIALS & CONSTITUTION

RECYCLED & READAPTED

MODULATED & PRE-ARRANGED

DIGITAL FAB

BERTINA, Stéphanie

Agence PCA

stbertina@hotmail.com

In the Basic Unit House [BUH], peruvian traditional knowledge of building in straw is revised with the input of modern technologies. Materials [straw and pallets] are available on site. The [BUH] is an easy building habitation. The wood structure is simple and respect the environment. Pallets are used to maintain the straw. A coat of lime is laid inside to protect the house. To facilitate construction, dimensions are relatively small: 6.47m x 2.87m x 4.24m. Considering the hot climate, the [BUH] is designed with small openings. The roof protects inhabitants from sun and casual rain. The terrace is north orientated to enjoy maximum sunshine. Rooms are located in the south fresh part of the [BUH]. The [BUH] is raised - built on piles - to respect natural ground and be environmental friendly.

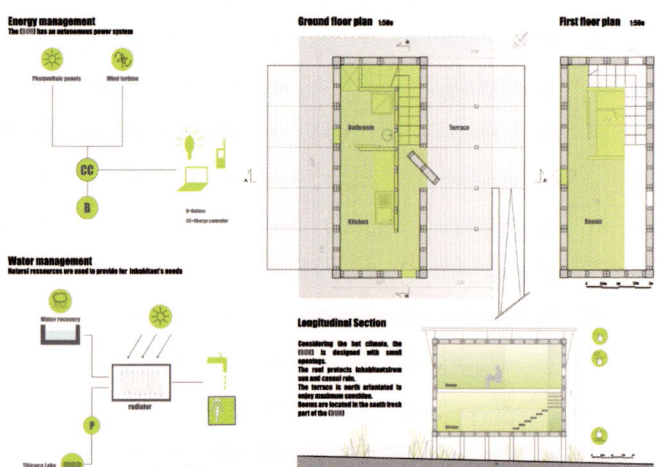

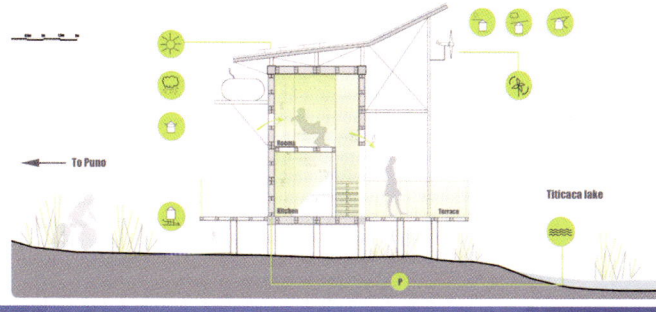

To Puno

Rooms

Kitchen

Terrace

Titicaca lake

DO IT YOURSELF

AUTO- & BIOGENERATED

MATERIALS & CONSTITUTION

RECYCLED & READAPTED

MODULATED & PRE-ARRANGED

DIGITAL FAB

BARSAN PIPU, Claudiu
DRAGAN, Andrei
NITUICA, Oana Maria

UAUIM

barsanpipu@yahoo.com

Our proposal tries to answer the most important needs of everyday life in a new and environmentallly friendly way. Tries to help the gypsies use their own "know-how" as long as their way of living in combination with some ecological affordable approaches. The main goal is to create a mobile habitat that can be built and adapted to the nomad lifestyle as fast as often as it is needed. We emphasize on the low-cost and low-tech solutions that can be often used, especially using the gypsies own indirect recycling habits. Now they can build their homes as they like to, in accordance to their esthetical culture but in a new and innovative way, using the "custom ECO bricks" that they can fabricate by themselves. We want to promote a new image, that of a GREEN GYPSY, away from their current negative stigmata.

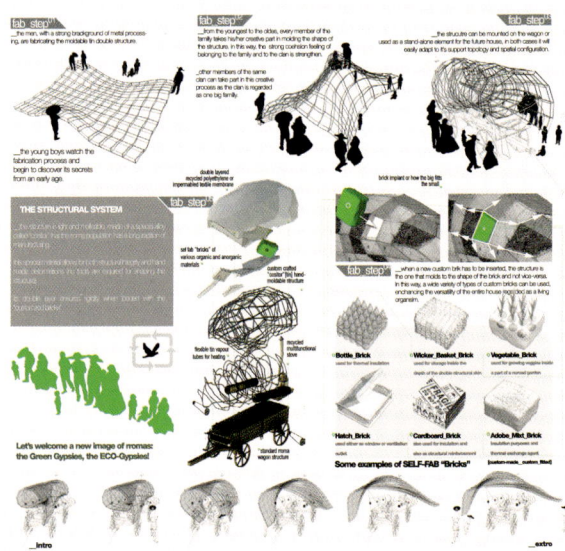

(alt. fab bricks)

DO IT YOURSELF

AUTO- & BIOGENERATED

MATERIALS & CONSTITUTION

RECYCLED & READAPTED

MODULATED & PRE-ARRANGED

DIGITAL FAB

MALAGA, Peter
BOECHAT CORDEIRO, Stella

BO MA+

malaga_peter@yahoo.com

Taking as an example the recently floods in south asia we decided to make a project for a slum community located on the banks of sabarmati river in central ahmadabad, where there are regular floods. We centralized on the Sunday market community, placed on the border between the future developed area and the left-overs of what used to be a natural environment and now is taken by the slums. SU market is historically a place for meetings and social activities, not only for the slum community. Now it is located in an area which is polluted and empty during the week and becomes alive on Sundays. We propose an artificial and natural overlapping. We redefined the sense of SU market by creating a kind of mixture of market, housing and landscaping. The area would be also a research village where the community could be informed how to build their houses by themselves. The basic feature of the project is the use of available resources, moreover the reuse of things which lost their primary utility; reused plastic barrels, containers, palletes, etc., combined with natural materials, such as bamboo. The plastic structure consists of joined barrels and stacked palletes, some of these barrels also work as water collectors. All plastic structurevis covered by plastic sheets and the bamboo structure consists of tied raw bamboo sticks and acts like natural "refridgerator", bringing fresh air into the house. These two parts are covered by textile and plastic fabrics which provide sun protection and create more intimacy.

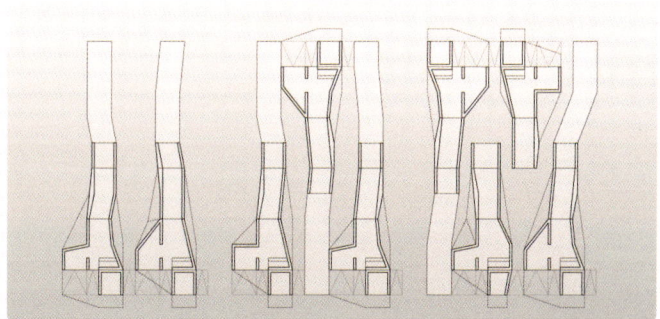

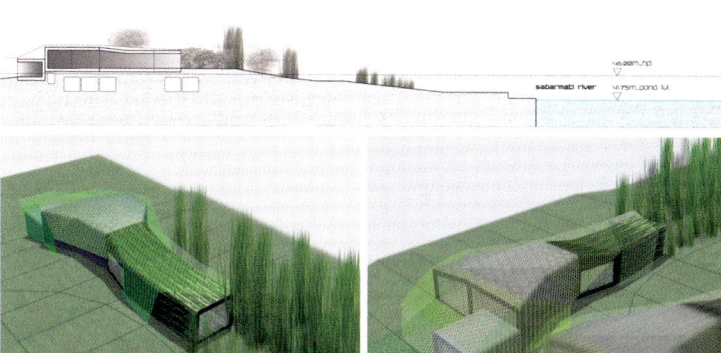

DO IT YOURSELF

AUTO- & BIOGENERATED

MATERIALS & CONSTITUTION

RECYCLED & READAPTED

MODULATED & PRE-ARRANGED

DIGITAL_FAB

LUCAC, Martin

Academy of Fine Arts and Design Bratislava

martinlukac79@gmail.com

GARDEN VARIETY project is located on the outskirts of Bratislava, capital city of slovakia - one of the post-communist countries in central Europe. The project explores the possibilities of grafting a small sustainable community into one of the garden colonies. Garden colonies were the communists concept of leisure, often characterized by problematic soil conditions, water contamination and air pollution due to hard industry factories neighboring these areas. The project responds to the existing conditions and tries to reactivate these fringe areas by inserting hybrid typologies. The project consists of two interlocking parts: the house and the green house. Housing types have the appropriate form of geologic fault models, as if they were some sort of architectural ready-made. While the forms deal with geology and soil strata, the greenhouses explore the hydroponic and aeroponic techniques of growing plants without the use of soil. This project examines the double potential and different application of construction materials such as thermal insulation, which can be used not only for insulating the walls, but also as a growing substratum in the hydroponics drip irrigation technique. The hydroponic system, supplied with water from retention and filtration from rain, is operated by a computer and it can provide with crops and create a greenhouse effect for the living space below. This space is founded on a space-frame anchored to the concrete slabs, uses the passive heating elements such as heat absorbing materials in the floor wall and ceiling.

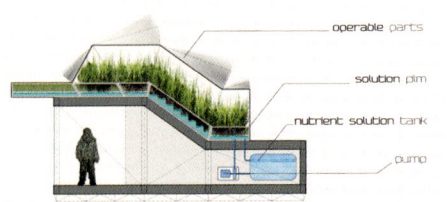

operable parts

solution pim

nutrient solution tank

pump

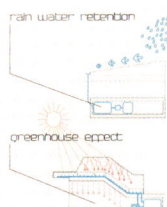

rain water retention

greenhouse effect

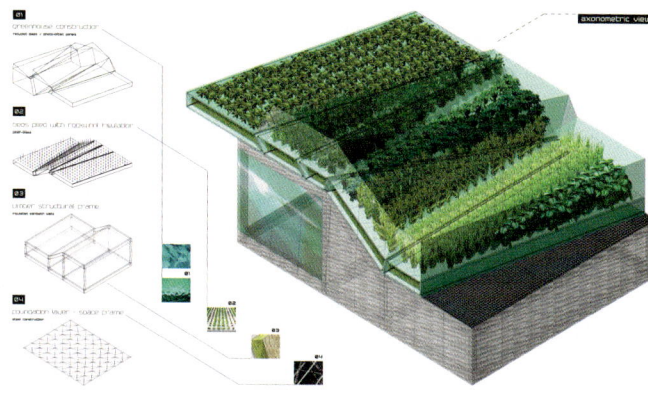

01
GREENHOUSE CONSTRUCTION

02
GREEN CARD WITH DECAYING VEGETABLES

03
UNDER STRUCTURE OF STONE

04
STRAWBERRY PALMS - SOLAR STONE

axonometric view

DO IT YOURSELF

AUTO- & BIOGENERATED

MATERIALS & CONSTITUTION

RECYCLED & READAPTED

MODULATED & PRE-ARRANGED

DIGITAL FAB

TANG, Ming
DIHUA, Yang

Architecture Department,
Savannah College of Art and Design

mtang@scad.edu

1ST PRIZE

The central feature of our bamboo house is the development of a deformable structure that exhibits characteristics of umbrella, with the potential of arranging themselves into various contexts and dwelling requirements. We named it as Folded House, a self reconstructive structure for instant installations, which, according to the changing internal requirements and site topography, can produce potentially infinite scenarios. Rather than using the industry mass production to generate uniform dwellings, the Folded House uses a simple kinetic structure made by bamboo, a kind of bottom-up assembling of complex adaptive systems that self-regulate, in opposition to top-down overarching principles. The straight bamboo poles are used to create ruled surfaces-helicoid, hyperbolic paraholoid and hyperboloid of revolution. The result of Folded House is a reflection of the logic of fold versus unfolds, self-construction versus de-construction, permanent structure versus mobility.

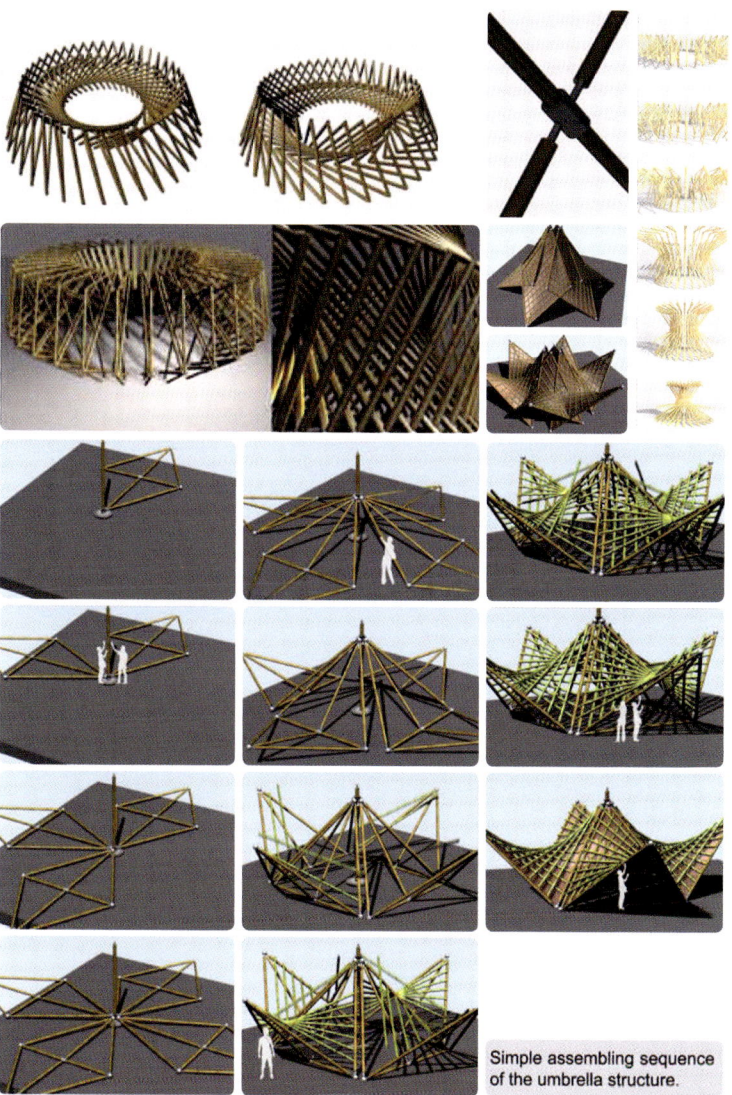

Simple assembling sequence
of the umbrella structure.

DO IT YOURSELF

AUTO- & BIOGENERATED

MATERIALS & CONSTITUTION

RECYCLED & READAPTED

MODULATED & PRE-ARRANGED

DIGITAL FAB

DO IT YOURSELF

AUTO- & BIOGENERATED

MATERIALS & CONSTITUTION

RECYCLED & READAPTED

MODULATED & PRE-ARRANGED

DIGITAL FAB

BEDU, Olivier
GESCHVINDERMANN, Christian
NORMAND, Sébastien
GABREAU, Marie

Cabanon Vertical

cabanon.vertical@laposte.net

Prototype

Location: directly in the tip

Size: until 5m high, 5m circumference

The annual production of used tyres in France was estimated in 2002 at 390000 tonnes. A tyre weights 10 kg on average. That is about 40 millions of tyres per year. How many of those are there in Vitrolles? A couple of millions? Anyhow according to our estimations we shall only need about 400 tyres to construct our "borie". A small corner of natured stuck between the road and a dump: we are definitely far from Luberon and its agragian plots. Nonetheless, having delimited a circle of tyres on the ground we reiterate the original gestures necessary to the elaboration of this building. Rank after rank, the circle narrows and by a simple stacking up in cantilever the ribbed vault takes shape. This "bad hat" (meaning of "borie"in Provence´s dialect) results from the haphazard meeting of a banal waste product and a constructive system inherited from the Neolythic.

In process

DO IT YOURSELF

AUTO- & BIOGENERATED

MATERIALS & CONSTITUTION

RECYCLED & READAPTED

MODULATED & PRE-ARRANGED

DIGITAL FAB

**RAGGI DE MARINI, Fiammeta
ROTA, Simone**

FACTORY3STUDIO

fiammettaraggi@yahoo.it

The aim of the project is the re-use of worn-out tires, in order to reduce their environmental impact and the problem of land fill sites. The poliuretanic foam conferes lightness and good insulation. It has great elastic qualities, it stands the test of time, cheap, biologically stable, non-toxic and it doesn't harm the environment. The ancient italian form of the trullo is the inspiration for both the structural and typological layout of the scheme.

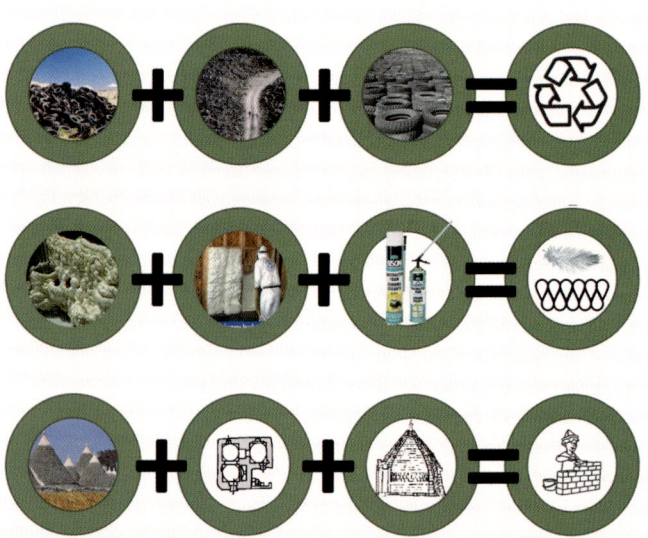

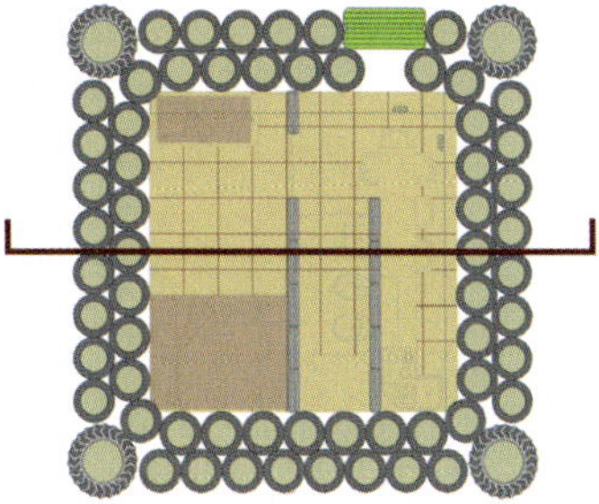

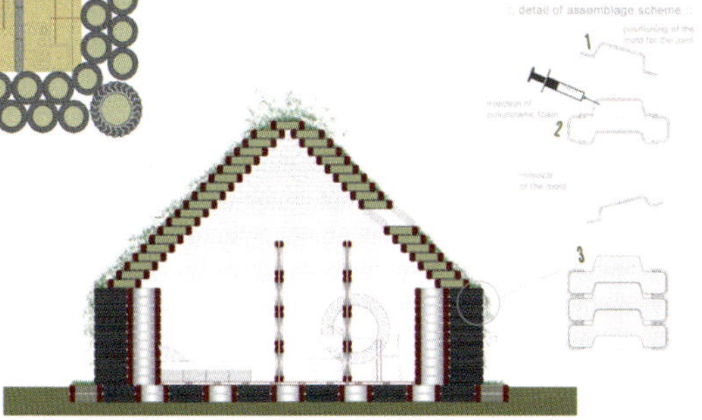

.. detail of assemblage scheme ..

1. positioning of the mold for the joint

2. insertion of polyurethanic foam injection

3. removal of the mold

DO IT YOURSELF

AUTO- & BIOGENERATED

MATERIALS & CONSTITUTION

RECYCLED & READAPTED

MODULATED & PRE-ARRANGED

DIGITAL FAB

MODULATED AND PREARRANGED

The idea of a modular element should be explored not only by studying particularly one module, but also by observing the way a number of modules participate in one structure. The structure itself is organized by specific and similar objects or elements that are repeatedly multiplied.

Consequently the construction industry is simplified, due to the fact that there are needed only specific "smart" objects (parts of the construction) that by multiplying them the whole structure is done. The extension of the construction is something again easily made, due to the fact that it can be done only by adding the same modular elements.

Their assembly is considered to be extremely easy and less time-consuming without the need of specialized labour; thus the construction is economically more sufficient.

There have been some intentions to create particular modular objects that would result in the maximization of the living space of the building.

Moreover, in some of the cases presented, there have been proposed modules that would have more effective result in the way the structure itself is behaving. On the other hand, some module's shape has been defined using the structural system as a basis.

Apart from the "smart" shape of a module either for its easier assembly, or for its better structural behaviour, there have been presented modules with "smart" and efficient properties. Modules can be made by specific materials; this would have a strong effect on their better adaptation to the environment and the changing climate conditions. Additional properties and elements would result in a more sustainable function of the module and consequently of the building as a whole.

Customization of modular structures could be studied in many aspects. Their "smart" shape could result in different combinations, depending on the personal preferences and needs of the user. What is more, the module itself could be reconfigured and change parametrically, according to the needs of the user or the characteristics of the area where is to be built. In this case, a specific joint system is needed so as for the parametrically modified modular elements to be connected.

DO IT YOURSELF

AUTO- & BIOGENERATED

MATERIALS & CONSTITUTION

RECYCLED & READAPTED

MODULATED & PRE-ARRANGED

DIGITAL FAB

ANTON, Juan Carlos
VILLA, Victor Ivan

Teoria Arquitectura

arquitecttheory@hotmail.com

National Meteorological System has reported that in Mexico more people die during the winter season than during the hurricane and rain season, although the fact that cold weather is not that extreme as in northern countries. In most of the cases people have accidents trying to protect themselves from cold weather creating bonfires or using electric heaters which are extremely dangerous, this becomes even more severe considering the living conditions of these people; spaces constructed by materials found by chance, such as cardboard boxes, metallic sheets or springs from old matresses. The government tends to ignore these situations, focusing on the cities, while the people themselves are trying to get away from them due to the fact that they cannot afford a piece of land there. This is not only a political situation, but also an architectural problem which should be solved. These garbage built spaces are pure architecture, people are creating what they need by what they have. It is our responsability to understand and learn from what these people need and what they do.

The main aim of the project was to maximise the space of a single structure without the need to add extra elements to the basic and simple form. Moreover, this project tries to reprogram the space in a group of growing modules reinterpreting the way of life in the region. By adapting the configuration of the traditional program, a simple box is turned with its properties into an antiprism which gives the possibility of maintaining the volumetric space of the box, enhancing the vertical connecting structure that is conformed by eight regular equilateral triangles.

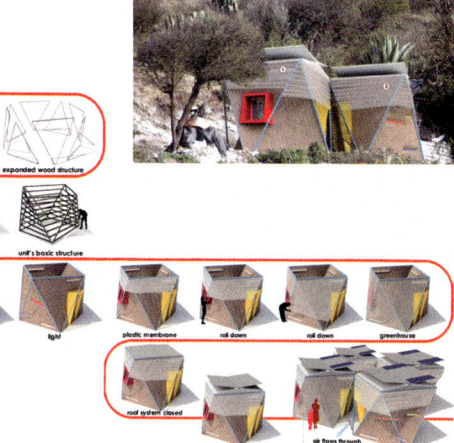

sun house

| 27 cubic meter space | maximize space on one structure | unit | expanded wood structure |

| structure | expanded wood wall structure | wall structure | unit's basic structure |

| adobe | access | ventilation | light |

| plastic membrane | roll down | roll down | greenhouse |

roof system closed

roof system open

air flows through lattice branch door
the membrane creates a greenhouse space facing south cycling cold air out and warm in.

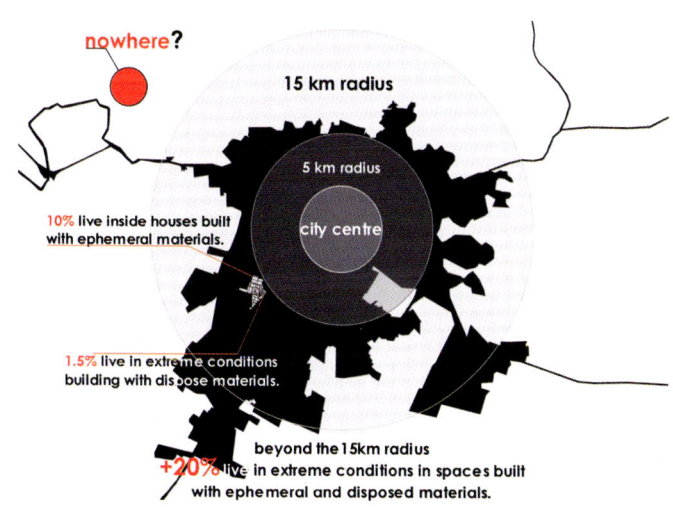

nowhere?

15 km radius

5 km radius

city centre

10% live inside houses built with ephemeral materials.

1.5% live in extreme conditions building with dispose materials.

+20% live in extreme conditions in spaces built with ephemeral and disposed materials.
beyond the 15km radius

DO IT YOURSELF

AUTO- & BIOGENERATED

MATERIALS & CONSTITUTION

RECYCLED & READAPTED

MODULATED & PRE-ARRANGED

DIGITAL FAB

OROZCO, Melisa
NAVARRO, Karla

ITESM

lepixlaxoury@hotmail.com

The year 2008 marks the first time in human history that 50% of the human population will be living in cities and the majority of that population shift is happening in rapidly developing countries such as China and India. The populations and the standard of living are increasing, these also puts a strain on the cities due to the fact that new housing and infrastructure are needed. How we deal with today's problems will have a direct impact on the way we live tomorrow.

The purpose of this project is to radically change the way we think about traditional housing conditions in cities. Housing does not have to be a burden to a city's development, but has the potential to form part of the solutions towards a sustainable future. Now is the perfect time to build the prototype. The Energy Tower in Chongqing is one of China's rapidly developing megacities who is struggling to keep up with housing and infrastructure demands. The Energy Tower has its own power, recycled water, waste management and air filtration systems that will service the tower's residents, who in return purchase prefabricated living and service modules and "plug-in" to the system. Eventually, more and more Energy Towers will be built to form a networked system of housing that will produce surplus energy to service the city, in effect creating a sustainable and near zero emission city.

This project is about a house made for two scientists living on a dry lagoon. The module is fully adaptable to all climatic change all year round and designed for the ever-changing climates. It is placed in the Sayula Lagoon.

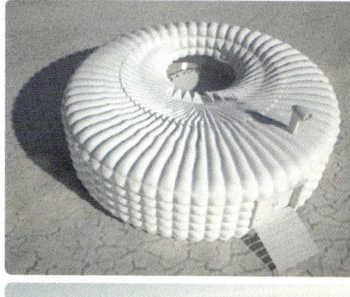

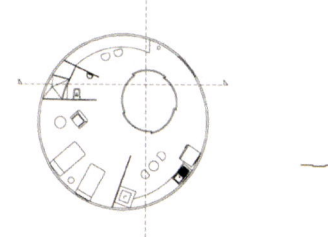

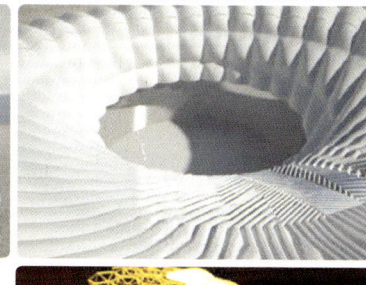

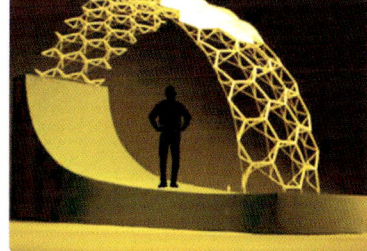

DO IT YOURSELF

AUTO- & BIOGENERATED

MATERIALS & CONSTITUTION

RECYCLED & READAPTED

MODULATED & PRE-ARRANGED

DIGITAL FAB

BARTESAGHI KOC, Carlos
GUILLEN DE ARCE, Dora
ZANABRIA OJEDA, Ulrich

Universidad Nacional de San Agustin de Arequipa

carlosbarta3@hotmail.com

The Chili River Valley is located between extensives urban degraded areas with a descontrolled growth, a real threat for its destruction, due to the fact that the city has been showing little interest to the river. The location presents two significant activities; agriculture and tourism.

The main strategy merges these work sources like a unique organism, inside of an extend net of interdependent symbiotic nodes (HBMS), applicable to other agricultural zones in Peru.

So the CHACARERO (urban farmer) and the tourist work together in all phases of development (planning, design, self-manufacture and assembly of the model kit). The building is conceived as an extension of the landcrops and is made with technics, materials and tools taken from the local context.

The structure and envelopes are 100% recycled and reused, and changes according to climate variations or number of occupants. All spaces change with the time according to new needs of their inhabitants. The entire structure is a big system of heating and cooling like a biological radiator with 100% of Lo-tech performance.

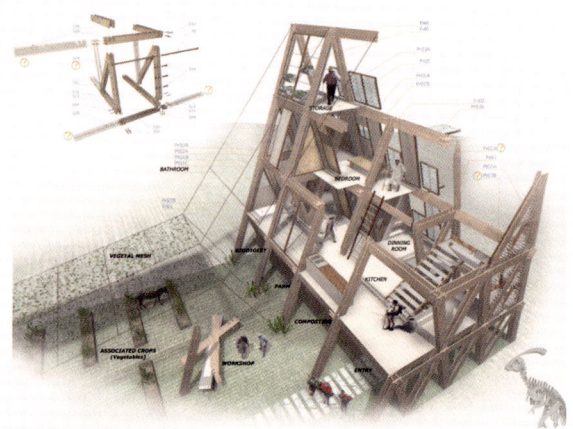

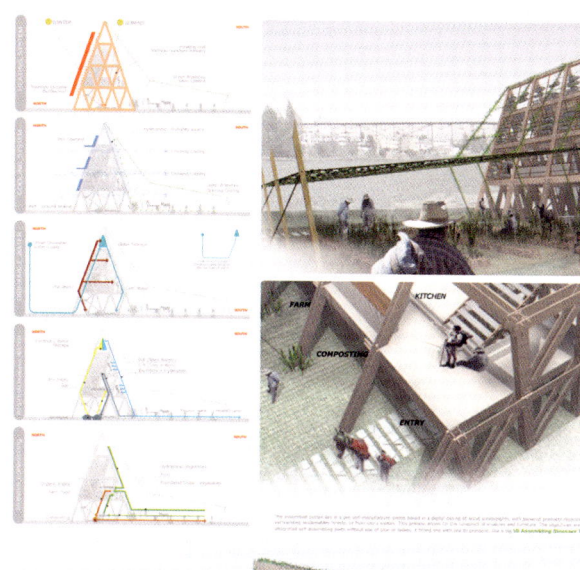

FARM

KITCHEN

COMPOSTING

ENTRY

DO IT YOURSELF

AUTO- & BIOGENERATED

MATERIALS & CONSTITUTION

RECYCLED & READAPTED

MODULATED & PRE-ARRANGED

DIGITAL FAB

DADOK, Pawel
ROGOWSKI, Michał

Silesia University of Technology

paweldadok@wp.pl

The idea of our project was to create a building which man could quickly build by himself, as well as pull it down and relocate it. Due to this idea we used a triangle which is a statistically unchangeable shape. As a result is constructionally independent. The construction consists of light pre-fabricated elements which can be transported and assembled easily. A frame of the network is filled with triangular elements which are able to montage according to self intention.

Due to its constructional system it adjusts itself due to one´s specific individual needs. A very important fact is the ability offered for creative actions to be expressed in order to adjust a block of the building. It is achievable thanks to the changeable design of the elevation. The established modular system enables to determine the size of the building and freely shape its structure and functional set-up. The elastic construction of the building provokes us to consider an architecture which is easy to built. It enables to raise not only the single building or conglomeration of these buildings but also to create more complicated structures which join themselves both perpendicularly and horizontally. The building, due to its universal form and various abilities of self-sufficiency is able to be located on almost every place of the world. Regardless of the climate and its resistant construction is able to endure regardless of rain, snow or earth-quake.

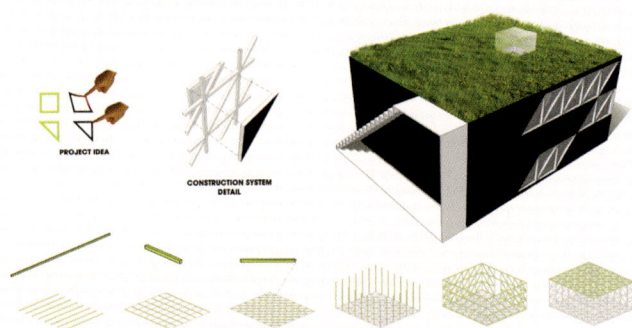

PROJECT IDEA

CONSTRUCTION SYSTEM DETAIL

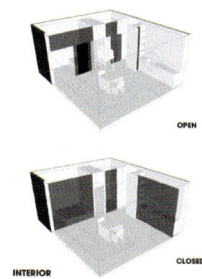

OPEN

INTERIOR   **CLOSED**

THE SELF - FAB HOUS3

DO IT YOURSELF

AUTO- & BIOGENERATED

MATERIALS & CONSTITUTION

RECYCLED & READAPTED

MODULATED & PRE-ARRANGED

DIGITAL FAB

DUARTE BENTO, Pedro

pedro duarte bento · genotype of primitive architecture

pedroduartebento@gmail.com

The Self-Fab H 4226 HOUSE (house for two to six people) is a prefabricated and energetically self-sufficient house. Its construction is based on the use of new digital technologies and is especially flexible to allow the owner design it and built it himself. With this project we are trying to break the traditional line between the buyer, the builder contractor and the real estate promoter. At the same time we are seeking solutions for the problematic of rural abandonment in the south of the Peninsula Iberica. The plea of the major cities as long with the climate changes and the abandonment of the rural practices are transforming the rural landscapes into deserts. Based in both purposes the SELF-FAB H 4226 HOUSE tries to be a common and reasonable option for those who pretend to leave the city and live in rural communities. Far from the traditional construction process, these housing units can be built at an affordable and less expensive price.

SELF-FAB CONCEPT: The principle of the house is generated by a section line; a contour line. The client can draw a section line that fits inside the measures of the cutting-pieces machinery. Once the contour line differs from one to another, the number of pieces is always the same and its way of assemblage has to be always the same.

EXPANDABLE CONCEPT: In order to obtain a larger flexibility from the users, the SELF-FAB H 4226 HOUSE was developed to allow its expandability. While the family grows, the house can grow along with it. With the Self-Fab files from the initial section line, more pieces can be fabricated and therefore expand several square meters of the actual living space.

SELF SUFFICIENT CONCEPT: The house is energetically supported by eolic and and solar energies. The water supply is guaranteed by the water reservoirs which storage rain water. The domestic water is filtered by an osmosis filter and the non-filtered water is used for irrigation of the vegetable garden and for the livestock animal care. The larger amount of energy comes from the photovoltaic skin.

COMMUNITY: In a self-fab community the users can share, and therefore improve, common equipment and tasks. Beside the technical equipment that provides self-sufficient resources to their own houses, the community members can add more or maintain gardens.

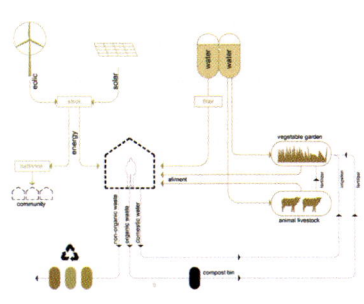

DO IT YOURSELF

AUTO- & BIOGENERATED

MATERIALS & CONSTITUTION

RECYCLED & READAPTED

MODULATED & PRE-ARRANGED

DIGITAL FAB

ASSEMBLAGE PROCESS FOR CASE [H44]
the five steps to assemble the house for four

design your house, build it yourself

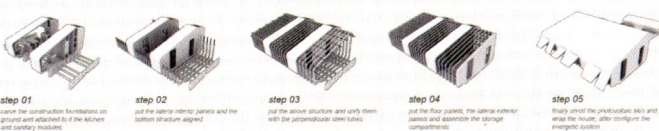

step 01
carve the construction foundations on ground and attached to it the kitchen and sanitary modules

step 02
put the lateral interior panels and the bottom structure aligned

step 03
put the above structure and unify them with the perpendicular steel tubes

step 04
put the floor panels, the lateral exterior panels and assemble the storage compartments

step 05
finally unroll the photovoltaic Skin and wrap the house, after configure the energetic system

ASSEMBLAGE PIECES

wind turbine
photovoltaic skin
kitchen module
sanitary module
piece A1
piece A2
pieces C1 and C2
piece B1
piece B2
piece D

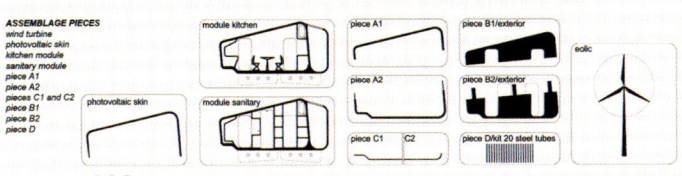

module kitchen

photovoltaic skin

module sanitary

piece A1

piece A2

piece C1   C2

piece B1/exterior

piece B2/exterior

piece D/kit 20 steel tubes

eolic

EXTENSION PLAN FOR H::4226
(HOUSE FOR TWO TO SIX)

and expand it as your family grows

YEAR 2007 - HUSBAND AND WIFE
HOUSE FOR TWO [H42]
AREA 41.44 m2

YEAR 2009 - THE COUPLE AND TWO CHILDREN
HOUSE FOR FOUR [H44]
AREA 63.56 m2

YEAR 2014 - THE COUPLE AND FOR CHILDREN
HOUSE FOR SIX [H46]
AREA 95.54 m2

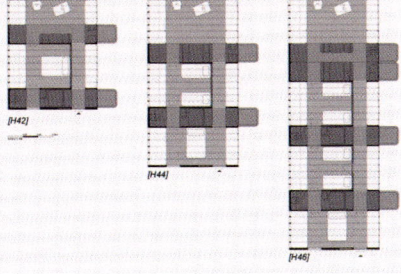

[H42]

[H44]

[H46]

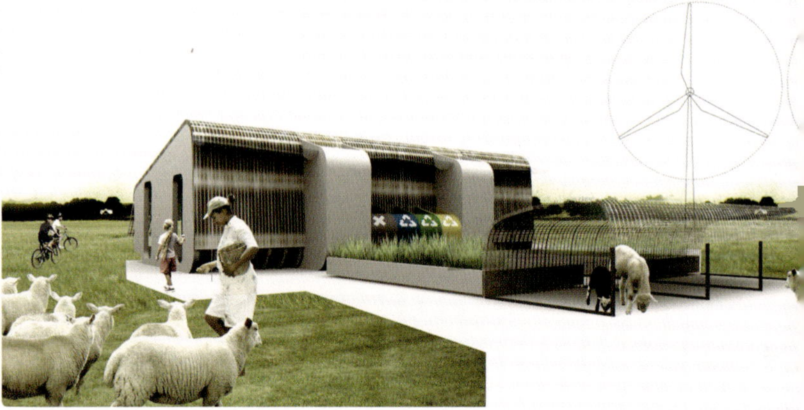

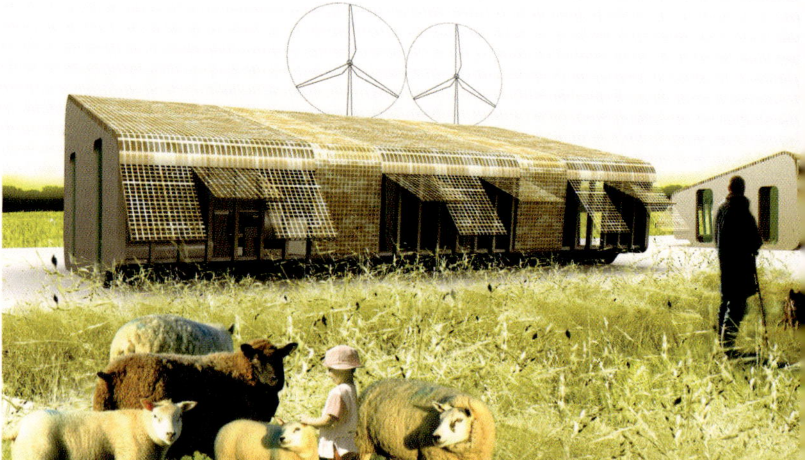

DO IT YOURSELF

AUTO- & BIOGENERATED

MATERIALS & CONSTITUTION

RECYCLED & READAPTED

MODULATED & PRE-ARRANGED

DIGITAL FAB

SALLOUM, Tarek

Dar El Handassah

tareksalloum@yahoo.com

"The following architectural concept routs away from a usual image of an architectural object and transcend a possible example of using nontraditional architectural technologies for the purposes of an architectural construction." The essence of this concept is the creation of a multi-functional interactive architectural space, also a space able to change its geometrical configuration depending on usage demands. Personalizing the habitat is important so as to achieve the user´s self satisfaction configurations. Constructively, the structure appears as a synthesis of hydraulic technologies and experience of membrane structures. The whole object is an umbrella system of three main transformable loops. Each loop represents a kind of transformable element, to which a system of contractible ribs and girders carcass for membrane cover are connected. It is a breathing system which once it is turned on so as to accommodate the user it inflates and shrink in height. On the contrary, the system will shrink and close as the user leaves the livable cell.

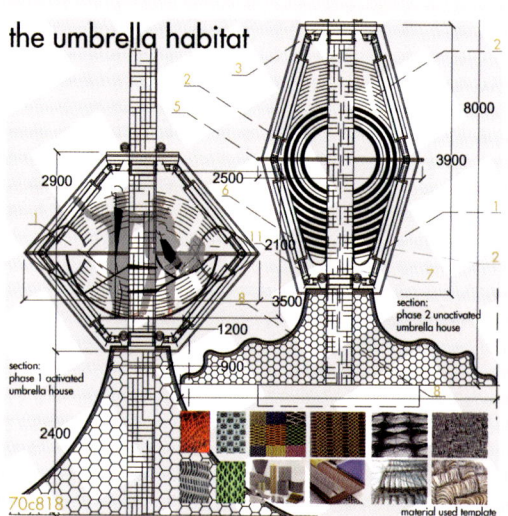

DO IT YOURSELF

AUTO- & BIOGENERATED

MATERIALS & CONSTITUTION

RECYCLED & READAPTED

MODULATED & PRE-ARRANGED

DIGITAL FAB

inner umbrella

KANTH, Chandra
VINOOB, S.

National Institute of Technology

archi_chandru@yahoo.com

The mechanized world has depleted most of the earth's valuable resources. A need for the use of renewable resources has taken shape. The cube can adapt to any climatic change and could be placed anywhere. The envelope of the cube is double layered concrete panel and the cavity between the concrete panels is used to carry service pipes. The solar panels are fixed to the outer concrete panel and are aligned to the direction of sun. One half of the roof has the solar panel which is inclined at an angle for the catchment of water and for the maximum efficiency of solar cells and the other half has the roof garden which utilizes the solid waste as compost. The plinth of the building is made of service chambers.

In short the building does the following functions:

— The energy needs of the building are satisfied by solar cells and by decomposition of soil waste (methane gas CH4).

— The building envelope acts towards the climate.

— Recycling of solid waste (compost).

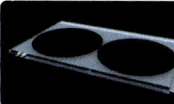

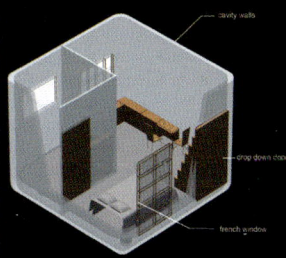

- cavity walls
- drop down door
- french window

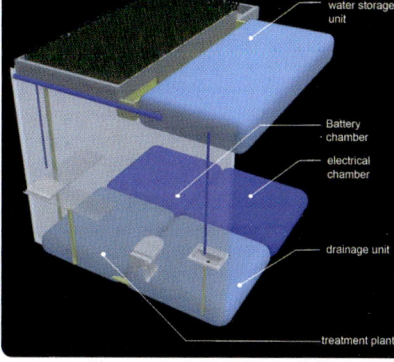

- water storage unit
- Battery chamber
- electrical chamber
- drainage unit
- treatment plant

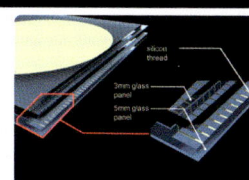

- silicon thread
- 3mm glass panel
- 5mm glass panel

DO IT YOURSELF

AUTO- & BIOGENERATED

MATERIALS & CONSTITUTION

RECYCLED & READAPTED

MODULATED & PRE-ARRANGED

DIGITAL FAB

CORY, Joseph
ADRIANI, Flavio
CARRY, David
FRIED, Eyal

Geotectura /Artecoop

geotectura@gmail.com

The Hydramid is a system of diamond-shaped pyramids designed to exist in off-shore waters. It is a rapid, semi-permanent housing solution for a populous of hundreds in a remote location, where land is scarce or hazardous, or where natural catastrophe has struck. Manufactured on-location, they are composed nylon form for structure, polyrethanic foam for filling. Despite its blunt material artificiality, the Hydramids blend, almost seamlessly, into the local ecosystem. The Hydramid diamond is cut in half at sea level. In its internal design, the top, floating half of the structure, serves domestic living functionalities and is divided accordingly. The bottom, underwater half, is used primarily as desalinized water storage space, but also for recreational activities such as marine-life observation. Upon launch, each Hydramid is uniquely Idled and enters the computerized HydraGrid, finding its exact location using the GPS system with mobility power supplied by a propelled motor.   The Hydramid recycles solar, wind and wave energies into electrical power, ventilation and water desalination. These resources are not dedicated to human living alone; other living agents such as greenhouse plantations and coral colonies feed off the energy cycle as well.

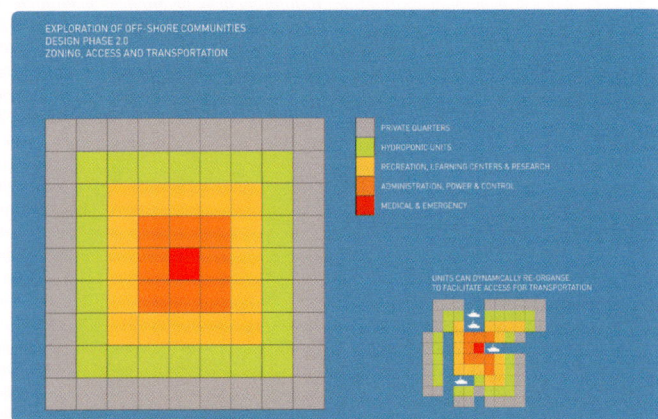

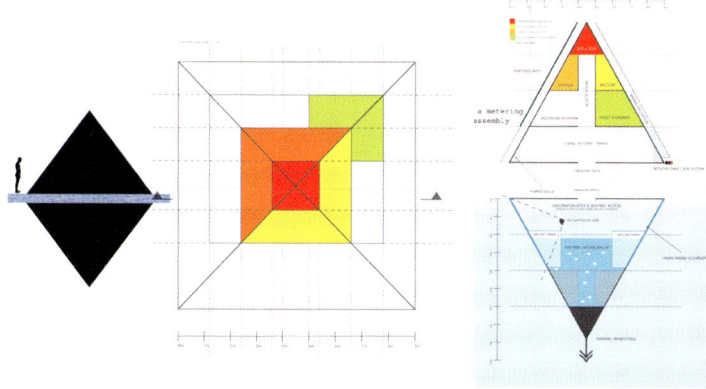

a metering
assembly

DO IT YOURSELF

AUTO- & BIOGENERATED

MATERIALS & CONSTITUTION

RECYCLED & READAPTED

MODULATED & PRE-ARRANGED

DIGITAL FAB

JUAREZ, Alberto   Universidad Iberoamericana

alberto.juarez.lucio@gmail.com

Like a magic box, this universal dwelling opens, folds and interacts in a simultaneous act for a sequence of possibilities.

An itinerant object that follows people on the move, it can be used from the city to the desert, passing through the forest. For people that want to extend their home and bring it with them, or for those who will use it as a permanent house, becoming a progressive growing system of living.

It is open to different possibilities of uses, either as a single unit or it can duplicate as part of a more complex system of organization. I imagine it as part of a small community, also pulled by a car that drives around the country or placed between two buildings in an urban environment.

The interior surfaces interact with the uses. The floor becomes a table, the bed is stored inside a wall, and the mechanical instruments are tightly placed.

As part of the strategy, the project includes a series of Self-Construction-Steps to assemble the object using the materials available in different conditions, becoming then, affordable for all.

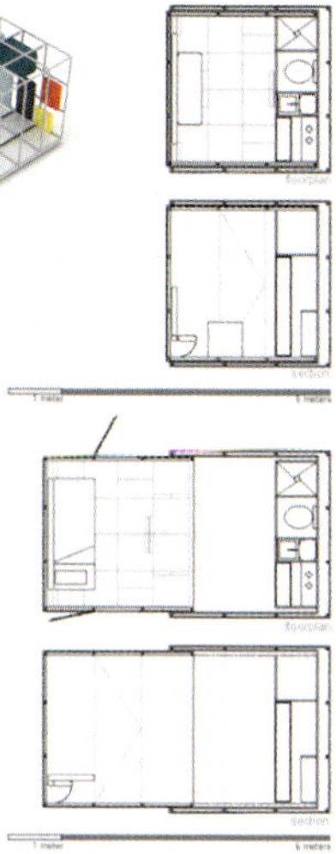

DO IT YOURSELF

AUTO- & BIOGENERATED

MATERIALS & CONSTITUTION

RECYCLED & READAPTED

MODULATED & PRE-ARRANGED

DIGITAL FAB

STEVAN, Luisa

IUAV Venice

luizaluigi@yahoo.it

"Fieldhouse" consists of a linear prototypical system using modular houses combined autonomously and according men's needs. These houses can be detached or be put in groups, creating a uniform surface. The house sustainability thus lies in matching its surroundings. The project is located in the widespread city of Veneto in Italy. Its shape can be applied everywhere in the world and it can't be visible by a satellite. In the future the houses will cover the big metropolis' buildings and make them become large, vertical fields.

The houses are made with two shapes: a self fab hexagon, sustainable and modular, and a plain. It's a strip that isn't rigidly linear, on the contrary it's moving making a ripple between the ground and the cultivation level.

The project starts from the idea that the self fab house lies on the chance of self providing building materials. The main materials of the construction come from the field. The primary construction materials are clay, straw and wood, built up as hexagon, shape as bees do.

The unit composing the fieldhouse can be assembled in various ways according to the family needs. The house has a module for the water, for the snow, for the animals, as a livestock, and for the fish. The batterybox is a box with organic and biological materials. Some of methane battery is put in these boxes without oxygen, they degrade them and leave methane, useful for the house to make energy. This box, with the fotovoltaic panels gives energy to the house while the wood keep the house warm.

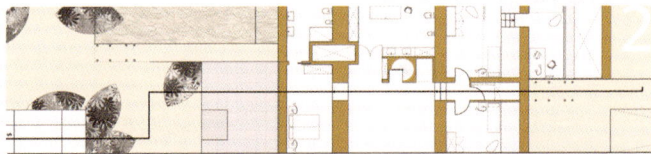

DO IT YOURSELF

AUTO- & BIOGENERATED

MATERIALS & CONSTITUTION

RECYCLED & READAPTED

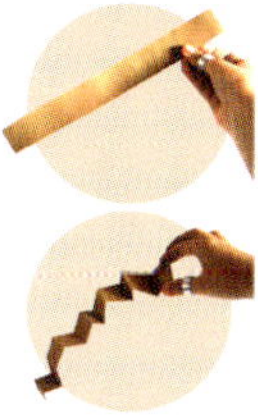

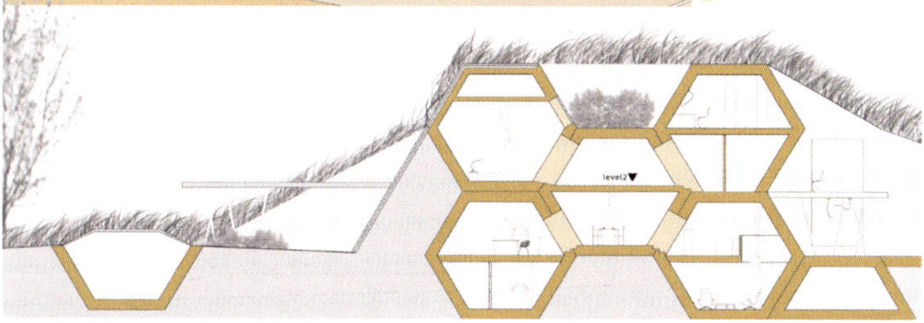

MODULATED & PRE-ARRANGED

DIGITAL FAB

THE CABINET HOUSE

YAMKLEEB, Jirawit

SCDA Architects

urbanomania@yahoo.com

THE CABINET HOUSE is a simplest module of daily life for a bachelor. At the present situation, when interiorization of urban space is omnipresent; more and more of public domains are covered and weatherproofed, the cabinet houses find themselves perfect sites for their colonization. Instead of adopting the conventional way of living, the cabinet house condenses all the necessary programs together and removes all internal connections. In other words, what we normally call circulation or corridor is flipped to the outside. By getting from one program to another, the resident has to step out of the house into public space first then get back inside again. In a day, as he or she continues his or her practice of daily life, the changes from private space to public domain, and vice versa, happens so many times that the frontier between public and private is dissolved. The space where the cabinet house situated has become part of the house. Adjacent public programs become extension of domestic uses. Every time the doors are opened, the programmatic shape of the house is reformed. The overall plan of this house is stretched as far as you can draw.

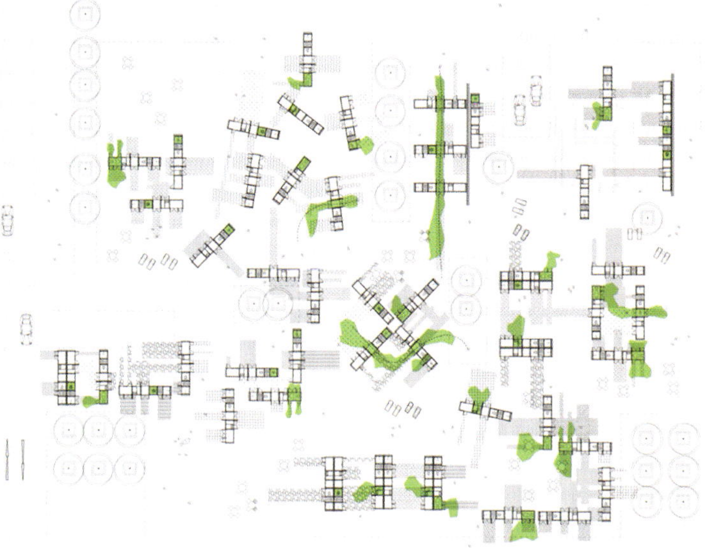

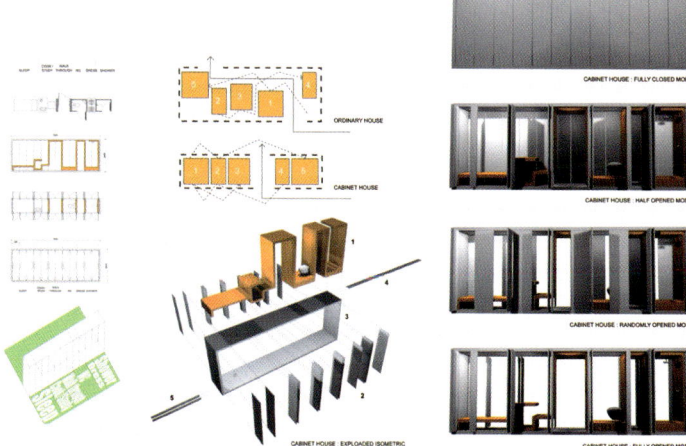

ORDINARY HOUSE

CABINET HOUSE

CABINET HOUSE : EXPLODED ISOMETRIC

CABINET HOUSE : FULLY CLOSED MODE

CABINET HOUSE : HALF OPENED MODE

CABINET HOUSE : RANDOMLY OPENED MODE

CABINET HOUSE : FULLY OPENED MODE

DO IT YOURSELF

AUTO- & BIOGENERATED

MATERIALS & CONSTITUTION

RECYCLED & READAPTED

MODULATED & PRE-ARRANGED

DIGITAL FAB

JITTAKASEM, Suebsai   Baan Khaophanon

                                     chittakasem@yahoo.co.th

The expansion of the city and the increase of the number of residences will result in the expansion of the transportation and communication networks. Although the population and its requirements are increasing, the resourses and the available areas themselves are in short. Thus, the aim of this proposal is to link a new architectural project with an older structure.

"UNDERSKYTRAIN HOUSE" is the concept house that links the communication, the living environment and the limit space of the city. This house can be attached to the skytrain rail structure. It should have compact size and be able to be adjusted due to the environment and the area that it will be added. The house placed under the skytrain rail has a main structure similar to the one of the rail itself . Staircases lead up to the house, which adjoints a collumn of the rail structure. The unit of the house is linked to the corridor under the house.

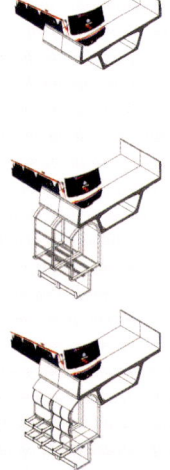

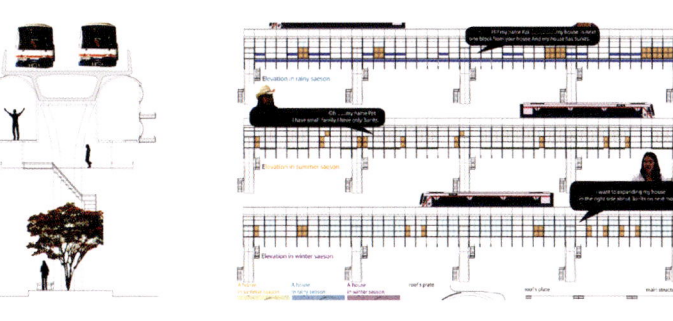

summer	rainy	winter
March - June	July - October	November - February

DO IT YOURSELF

AUTO- & BIOGENERATED

MATERIALS & CONSTITUTION

RECYCLED & READAPTED

MODULATED & PRE-ARRANGED

DIGITAL FAB

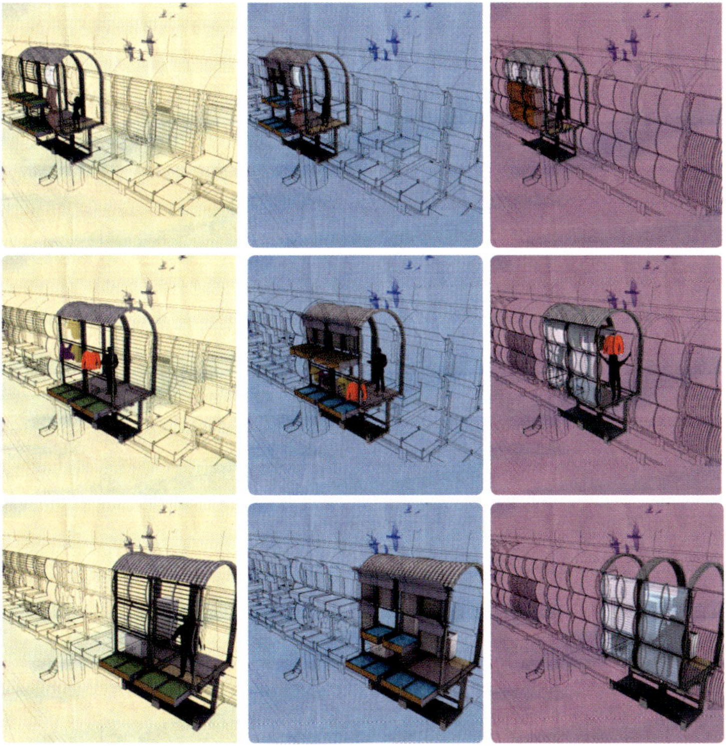

VANN, Andy
GEROUSSIS, Alexandros
PEDRINI, Federico

OURoffice

dropfour.andy@gmail.com

The excluded...

Not everyone is included in the advanced capitalist society, no matter how humanistic it may seem. Even in the presence of a sophisticated social welfare system self-fabrication and self-sufficiency become necessities to declare the presence of the most basic necessity: the right to self...

Self-sufficiency is radical!

Self-fabrication is liberating!

During copenhagen's cold and long winter, people outside of the traditional system need places of their own. Shelters can promote dependence. A self-fabricated shelter, made by the different elements of fringe society could create a small self-sufficient urban community where the forgotten reassert themselves. It can be environmentally and socially self-sufficient.

The excluded space...

Gaps in the urban fabric develop over time. Alleys, doorways, and parks are places where the excluded seek shelter. Our project finds one such niche in the city block.

Constructed from a simple system of scavenged elements (palettes and scaffolding members), the inserted structure aims not only to create a constructive environment, but also an exciting spatial experience. Hanging above street level provides a priveleged view, while the minimal structure and materials allow the protective space to be bathed in light and well-ventilated. In addition to providing a nurturing space to its inhabitants, the larger community's public space is enlarged with a street level community garden, cultivated in part by the by-products of life in the hanging community.

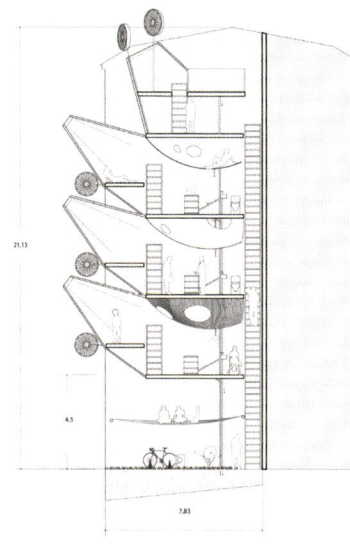

21.13

4.5

7.83

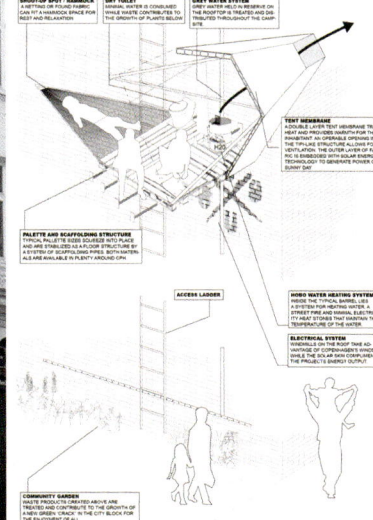

SHOOT-UP SPOT - HAMMOCK
A NETTING OR FOUND FABRIC CAN FIT A HAMMOCK SPACE FOR REST AND RELAXATION.

DRY TOILET
MINIMAL WATER IS CONSUMED WHILE WASTE CONTRIBUTES TO THE GROWTH OF PLANTS BELOW.

GREY WATER SYSTEM
GREY WATER HELD IN RESERVE ON THE ROOFTOP IS TREATED AND DISTRIBUTED THROUGHOUT THE CAMP-SITE.

TENT MEMBRANE
A DOUBLE LAYER TENT MEMBRANE TRAPS HEAT AND PROVIDES WARMTH FOR THE INHABITANT. AN OPERABLE OPENING IN THE TIP-LIKE STRUCTURE ALLOWS FOR VENTILATION. THE OUTER LAYER OF FABRIC IS EMBEDDED WITH SOLAR ENERGY TECHNOLOGY TO GENERATE POWER ON A SUNNY DAY.

PALETTE AND SCAFFOLDING STRUCTURE
TYPICAL PALLETTE SIZED SQUEEZED INTO PLACE AND ARE STABILIZED AS A FLOOR STRUCTURE IN A SYSTEM OF SCAFFOLDING PIPES. BOTH MATERIALS ARE AVAILABLE IN PLENTY AROUND CPH.

ACCESS LADDER

HOBO WATER HEATING SYSTEM
INSIDE THE TYPICAL BARREL LIES A SYSTEM FOR HEATING WATER. A STREET FIRE AND MINIMAL ELECTRICITY HEAT STORES THAT MAINTAIN THE TEMPERATURE OF THE WATER.

ELECTRICAL SYSTEM
WINDMILLS ON THE ROOF TAKE ADVANTAGE OF COPENHAGENS WINDS WHILE THE SOLAR SKIN COMPLEMENTS THE PROJECT'S ENERGY OUTPUT.

COMMUNITY GARDEN
WASTE PRODUCTS CREATED ABOVE ARE TREATED AND CONTRIBUTE TO THE GROWTH OF A NEW GREEN 'CRACK' IN THE CITY BLOCK FOR THE ENJOYMENT OF ALL.

DO IT YOURSELF

AUTO- & BIOGENERATED

MATERIALS & CONSTITUTION

RECYCLED & READAPTED

MODULATED & PRE-ARRANGED

DIGITAL FAB

KEERNS, Ryan
SMITH, Kristen

University of Pennsylvania

ryan.keerns@gmail.com

While the United States has long been a destination for immigrants seeking better job opportunities , many new immigrants are not given the opportunity to live the American dream. The ironic truth is that while more than one third of the estimated 20 million immigrants from Mexico and Latin America acquire jobs in the construction and manufacturing industries, less than half will ever be able to afford to own their own home. The Pollen House attempts to change this by creating an industry of premanufactured structural panels that could be easily assembled on site by these new immigrants. Each house would be customized by the potential family by adjusting parameters such as desired square footage, and number of levels and parametrically modified to respond to variations in site. The panels would then be formed and issued to the client on site, along with a set of instructions. Using only a few common tools, these panels could be assembled by the family as easy as a piece of furniture from Ikea. The interior can then be built out following a protocol of jointery to meet the programmatic divisions of the house.

The complex network of plant pollination deals not only with the plants in question, but also insects, their migratory patterns and local wind currents. This disciplinary multiplicity results in patterns of behavior similar to the complex network of global economics. As the flows of capital, goods and services increase, the populations supporting that effort are equally in flux. Like pollen in nature immigrants are providing an important part of the labor force, sustaining economic prosperity, despite the contested nature of the current political environment.

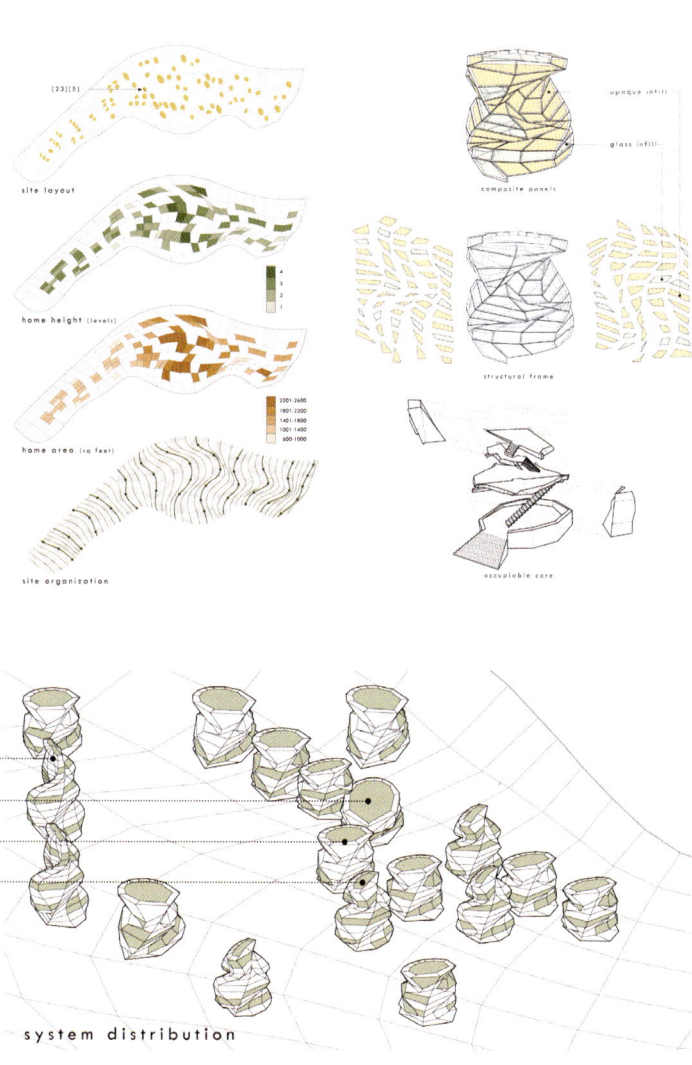

[23][5]

site layout

home height (levels)

4
3
2
1

home area (sq feet)

2201-2600
1801-2200
1401-1800
1001-1400
600-1000

site organization

composite panels

opaque infill

glass infill

structural frame

occupiable core

system distribution

DO IT YOURSELF

AUTO- & BIOGENERATED

MATERIALS & CONSTITUTION

RECYCLED & READAPTED

MODULATED & PRE-ARRANGED

DIGITAL FAB

TORRES, Luis

California Polytechnic University
l . a . T. d e S i GN iNC.

luisatorres@csupomona.edu

Mexico turned out to be a great project site due to its major need for better housing. It has also become a great provider of its materials and resources used for the proposed "nuevo milenio_ protype_ 01".

T. j. as we all know borders the southern part of the U.S.A. and the northern part of Mexico, due to this it has become a slum city for many. The majority of its inhabitants are dreaming of moving north searching for a better life.

What is proposed for this plot and as a prototype is a two-bedroom, kitchen, dining and living area home, that can be adjusted as needed. The sleeping areas share a common hallway that may lead to the additional bedroom, depending on how many the site allows us. This hallway is accompanied by an exterior corridor, its overhang, although it shades the house during the summer season, it does not prevent the winter sun from penetrating into the spaces. Moreover, the outdoor land is serving as playing area as long as common area for laundry. Another aspect of this proposal is the carport that is penetrated into the land and creates a deck space for entertainment and viewing.

The house is built by (Typ.) wood frame construction at the walls with recycled rags and clothes from the local factories used for insulation. The foundation and retaining walls are made with a gabion system that is filled with rocks from around the neighborhood. The outdoor hall of corridor would be made from recycled pallets and any steps necessary would be made from T.V. sets and/ or tires left at the local dumps. The roof will also include recycled pallets as shingles and recycled plywood pieces for sheating.

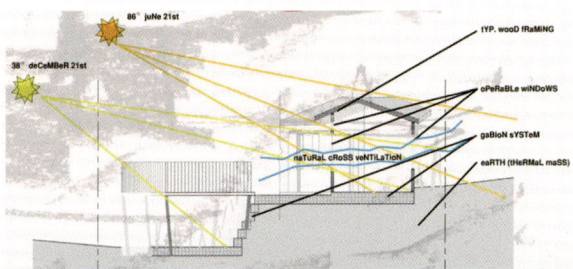

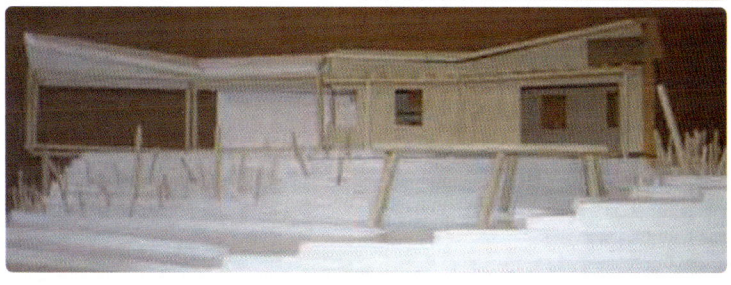

DO IT YOURSELF

AUTO- & BIOGENERATED

MATERIALS & CONSTITUTION

RECYCLED & READAPTED

MODULATED & PRE-ARRANGED

DIGITAL FAB

SANCHEZ RODRIGUEZ, Claudio   Udelar
BORGNO MARTIN, Javier Eduardo
                                              claumaro@gmail.com

We begin with the primitive idea of refuge, which has the function of offering protection from natural phenomena. From the beginning of humanity, the population has been changing its habits and its relationship with the environment, passing gradually from nomadism to the sedentary way of life. Nomadic population moved in the territory in search of natural resources that satisfied their consumption needs, and then they returned to move to another part in search of the same. The proposal comes from the idea of generating possibilities for the establishment of this kind of nomadic population. In some degree, It consists of the adaptation to the different and not adverse geographic situations with the minimum possible environmental impact. It is a basic house, self-contained and self-sufficient that is based on the land, which there is no need to be modified. The house lays down on the land by means of an hydraulic system. This system is made of a series of hydraulic action telescopic supports which adapt to the land situation in an automatic way. Once this stage is finished, the self-contained house unfolds in itself to give place to all the rooms. It is proposed a closing composed system by polymer layers to which air is injected, so as to make it harder and to give an air thickness between the layers. These layers improve the thermal behaviour of the house as a whole. By means of this device, bedroom and living-room are defined. The house is complemented with the integration of a photovoltaic panel system located on the cover, which assures a basic energy supply for the operation of the electric elements defined by the user.

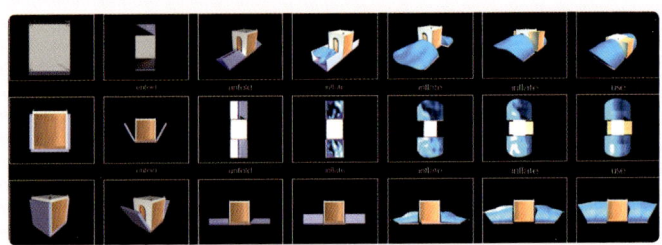

back view

left view

front view

floor plant

section

DO IT YOURSELF

AUTO- & BIOGENERATED

MATERIALS & CONSTITUTION

RECYCLED & READAPTED

MODULATED & PRE-ARRANGED

DIGITAL FAB

FIGUEROA, Juan
GOMEZ, Javier

nido, paisaje y arquitectura

sosjm@comcast.net

For mankind house is a filter between the world and their soul. This house has been developed to create a spiritual space as a part of its infrastructure. To do so we proposed an additional room to isolate a person from his or her environment, we call it the sanctuary of inner man, a space for meditation to find a high level of conscience perhaps and probably appreciation of his or her ecosystem. On the another hand this could be seen as an additional function that is questioning our way of living. This house is a prototype of a low-cost-dwelling", which could be built anywhere in a tropical weather. Its bamboo structure is resistant to earthquakes and its value in South America is around 6.000 euros. Lake Calima is one of the largest artificial lakes in Colombia, build in 1961 and occupying an area of 70 kl. The Malagana Indian culture has existed 1900 years ago in this region. They visioned about dwelling as a unit of the cosmos.

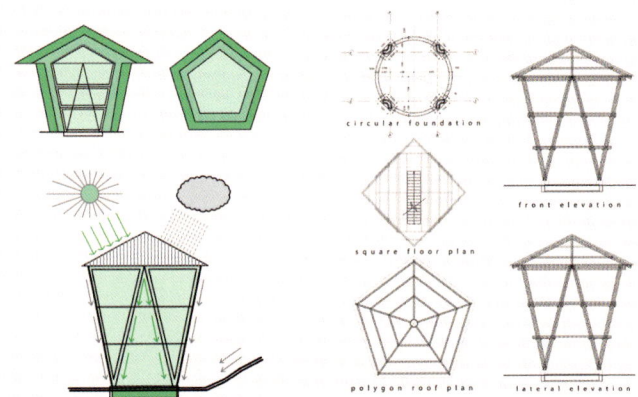

3 days - 3 people

Floor and roof

2 Deck, 2 stairs and 8 trusses

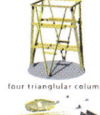

four trianglular columns

Structure by parts

one single assembly to build
the total structure

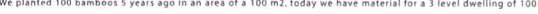

We planted 100 bamboos 5 years ago in an area of a 100 m2, today we have material for a 3 level dwelling of 100 m2

roof plan

front elevation

DO IT YOURSELF

AUTO- & BIOGENERATED

MATERIALS & CONSTITUTION

RECYCLED & READAPTED

MODULATED & PRE-ARRANGED

DIGITAL FAB

SCHUETTE, Oliver

A foundation

os@a-foundation.org

Intending to look at housing not as a problem, but as a generator of a sustainable urban environment, we introduced concepts and techniques for harvesting natural resources like rainwater, solar - and wind energy. These techniques ultimately start to shape the building proposals and its redefined urban development. At the same time, an avoided and sometimes frightening public space is being reformulated. Thinking of a society that is defined by continuous changes and collectively challenged to control its urban living environment, the proposals introduce living concepts such as 'work & live', the 'generational house', the 'growing house' and other typologies that can adapt to individual and collective life styles (another way of looking at sustainability).

The design proposals are images for a theory of change. Obviously, a realization of those would require more than what is usually included in the architects' portfolio. Some of the necessities might be a local or international infrastructure of financing that would see the potential of social housing as a motor for urban renovation. Until such mechanisms are installed (by the means of either local economy, private partner- or sponsorship, state responsibility, tax policy, international loans or subsidies), the task is to create visions for action.

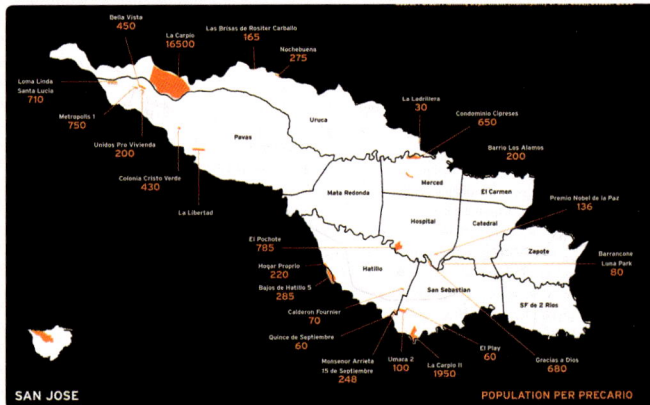

COLUMNSPACE - ...
EL POCHOTE

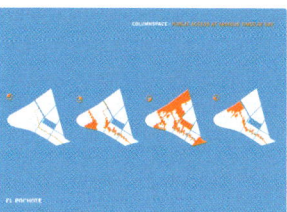

COLUMNSPACE - ...
EL POCHOTE

COLUMNSPACE - STREET PERSPECTIVE (OPEN FACADES)
EL POCHOTE

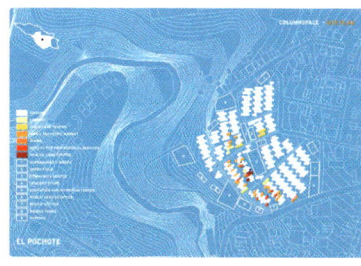

COLUMNSPACE - ...
EL POCHOTE

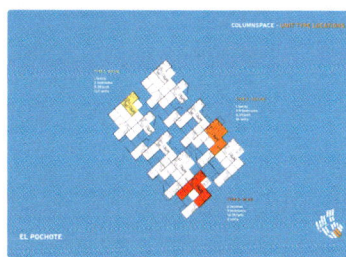

COLUMNSPACE - ...
EL POCHOTE

DO IT YOURSELF

AUTO- & BIOGENERATED

MATERIALS & CONSTITUTION

RECYCLED & READAPTED

MODULATED & PRE-ARRANGED

DIGITAL FAB

CHARRAT, Chrystel

Paul Andreu Architecte Paris

chrystel.charrat@gmail.com

The site proposed for the realization of this self-fabricated housing can be any forest. Ecosystems, are the shelters of million ground lives. They are threatened by numerous natural constructions and are varying both in their quantity and in their quality. It seemed natural to propose an auto-constructed house fitting into such an environment middle. Trees are necessary for the ground life, due to the fact that they provide it with natural shades and protect from the wind and the rain. Their more or less cowardly disposal allows to fit out naturally places of life. While observing the nature we learned that ants, little terrestrials, the spider, majestic tightrope walker, are the most significant sorts for constructing their domain of life.

Workers, as they have been observed, know how to take advantage of their environment and preserve it, by building just like the architects do. The monohouse wants to be the meeting of these two cultures of environment. Its architecture does not wish to produce just an object but to give a sense to the construction and to absurb it as much as possible in its environment: The monohouse is a work between the nature and the constructed.

DO IT YOURSELF

AUTO- & BIOGENERATED

MATERIALS & CONSTITUTION

RECYCLED & READAPTED

MODULATED & PRE-ARRANGED

DIGITAL FAB

DIGITALLY FABRICATED

Advanced Digital Technologies enable the possibilities of Fabrication. CAM Technologies give us the opportunity to produce physical things from virtual data. Technological Advances play an important role in our everyday life and can reconfigure the ways we are organizing our society as long as the relations and interconnections that we are establishing.

Digital Fabrication makes possible the rapid fabrication of the construction elements, additionally is economically more efficient.

Information age promotes the transmission and sharing of information. Designs themselves could carry information that could be shared in a platform. Each one who would have access to this platform could use and modify the uploaded designs depending on each own preferences and needs and finally fabricate them; due to the fact that the environmental conditions of an area as long as the climate ones, the user's characteristics, needs and preferences, could be translated into digital information themselves and then be applied as parameters that modify a specific building design. Customization is again strongly supported; the modified non-linear shapes are rapidly and economically fabricated.

Apart from the fabrication of a prototype or a series of structural parts, the whole building structure could be assembled by digital fabrication means. In this case the digital information of the exact position of each structural part is translated and it is applied in the construction site by robotic systems.

The new technological advances can give us the opportunity to explore new techniques, more sufficient ones, of both designing and production. They promote the experimentation on different materials, lighter and more efficient structural techniques and behaviors.

Digital Fabrication means promote the creativity and production freedom in the smaller scale; something completely opposite with what the big industrial systems do. The user himself could actively participate in the construction processes. As a result new relationships and interconnections during the production process are established.

DO IT YOURSELF

AUTO- & BIOGENERATED

MATERIALS & CONSTITUTION

RECYCLED & READAPTED

MODULATED & PRE-ARRANGED

DIGITAL FAB

BALASUBRAMANIAN, Prem
WINNIE I, Carola

National Institute of Technology
Tiruchirrappalli

prem_archi@yahoo.co.in

This house is made of rolled up nano polymer resin strips embedded with nanobots (quantum dots) enabling self construction and functioning due to the computing capability and integration of LEDs and solar chips for power and light. LEDs form the pixels of the house - making the entire surface to glow and display images and traces of skyline and camouflaging with buildings or surroundings. This enables virtual environment in the dwelling with any programmed images or background visuals. This house doesn't require any foundation making it possible to build anywhere from ground to terraces to a forest etc. The waterbed becomes the storage for collected rain water as well as supports the steel base grids to which programmed and flexible nano tubes are fixed. Following this the two layers of nano polymer resin are rolled on to the tube. Polymer resin structures are made like nerves of leaves allowing impregnation of nanobots - LEDs and solar chips and optic fibers to transfer data through out the surface of the house - making it an universal plug and play medium. All these make both assembling/ disassembling with minimal human intervention thanks to smart and intelligent building materials.

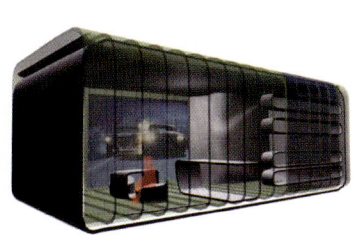

DO IT YOURSELF

AUTO- & BIOGENERATED

MATERIALS & CONSTITUTION

RECYCLED & READAPTED

MODULATED & PRE-ARRANGED

DIGITAL FAB

MADRIGAL, Maui
CONTRERAS, Ivon

Tecnologico de Monterrey Campus
Guadalajara

maui_87@hotmail.com

The Tuareg nomads are located in the vast areas of the Southern Sahara and they are characterized by their forte full independence. Deeply inmersed in a stunningly hazardous environment, they dedicate themselves to create beautifully abstract and pragmatic objects.

In the Tuareg culture the word for tent (mina or mana) is linked to the concept of marriage and the establishment of a new home. To them, marriage speaks of the unity of opposites that tend to a common goal. Each element is complementing the other and they are both creating a whole.

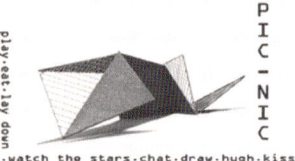

DAY

NIGHT

DO IT YOURSELF

AUTO- & BIOGENERATED

MATERIALS & CONSTITUTION

RECYCLED & READAPTED

MODULATED & PRE-ARRANGED

DIGITAL FAB

PANTOJA, Rodgrio
LOPEZ, Erbery
OCHOA, Levi
PALACIOS, Hector
PULIDO, Ruben
WELLER, Aaron

evO(a)_lAb

rod@evo-lab.com

A shipping pallet possesses temporal universality. It is a conductor of material exchange between all regions of this world.

It was once understood as a fruit bearing necessity, or a lintel at a threshold of a stone wall, or as a foundation for distributing products. At different moments and in different manners it has been scarce and abundant, 'pure' and modified.

We have envisioned the components of pallets to be appropriated and re-ordered. Each single, planar piece is cycled to compose a structural system of enclosure. A structural pallet system allows a range of diverse in-fill materials. A light-weight tensile membrane envelopes or a compressive earthen material fills in a structural pallet system. Region defines which material is most suitable.

Maybe structural pallets are combined with a woven fabric of hand-stitched plant fibers or 'worn-out' mechanized sweaters. Scrap or cut sheet or strip metals and plastics can also shroud pallets. These examples of light-weight tensile membranes are tucked and pinned in between two structural members. A heavy compressive earthen material simply fills in the square void of a pallet system. Blocks of adobe, brick or concrete can be stacked or laid with mortar. In addition, each square void can become a window – framed and inset with glass.

Needs and wants of contemporary inhabitants have inspired products beyond simple fabrication methods able to be utilized by an unskilled individual. Inside and at the extents of ceiling and floor we propose integrated systems; modules of components made in different regions, assembled in another location before located on site. Modules for water and energy collection and distribution, bathroom and kitchen services, technological gadgets, and furniture are complex composites – different from an in-situ pallet enclosure 'module' of immediate surroundings.

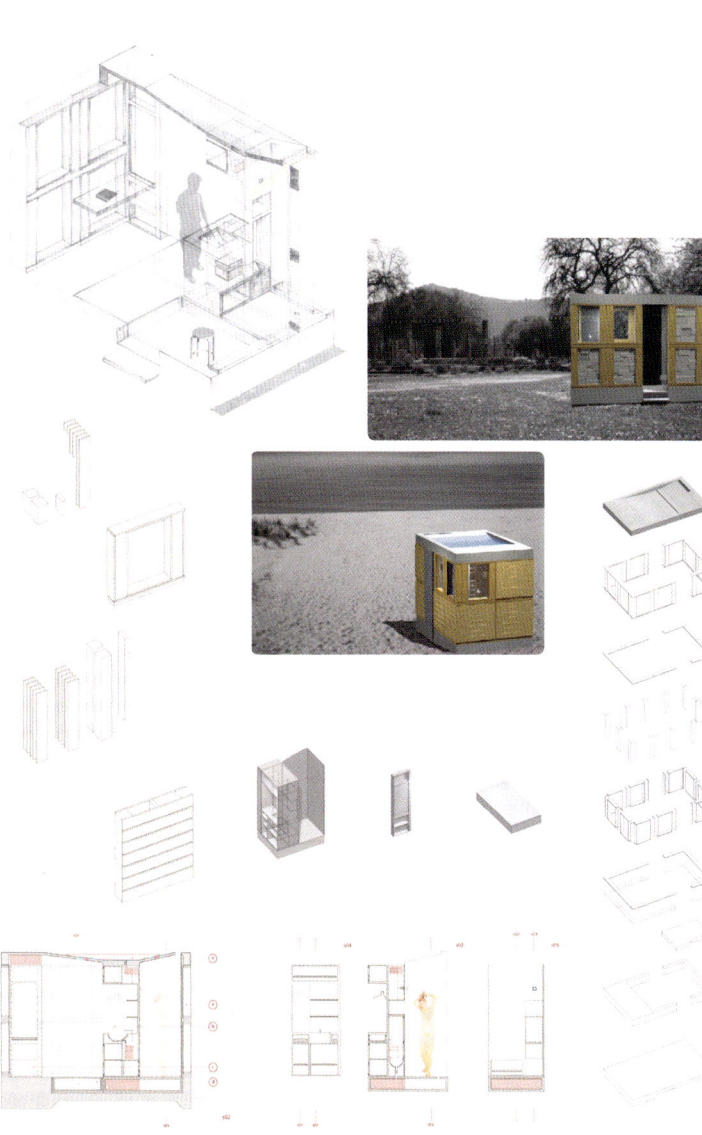

DO IT YOURSELF

AUTO - & BIOGENERATED

MATERIALS & CONSTITUTION

RECYCLED & READAPTED

MODULATED & PRE-ARRANGED

DIGITAL FAB

SEUNG TEAK, Lee   Harvard University
HYUNG JIN, Choi
KYUNG JAE, Ryu   sl2741@columbia.edu

The Alang Sosiya Ship breaking yard (ASSBY) along the coast line of India had commerced its operation on 1983 and became one of the largest station of such an industry due to its conditions; 17m tidal change and 25 m/s wind force.

However, the huge amount of workers at the yard are facing a variety of problems, especially the terrible condition of temporary living. Most of them live in small dwellings without sufficient infrastructures and potable water. Considering the process of ship breaking, their low income and many other aspects, there is a necessity of housing that can be easily manufactured and have its own infrastructure using the specific environment on the site. The components of ships and the characteristics of natural environments could be effectively utilized in the construction. This proposal will provide self-sufficient collective housing and infrastructures to the migrant workers and their families. Self-Sufficiency in this project does not only mean the housing's sustainability by harnessing natural energy, but also the continuous re-assembly of dwellings by the workers themselves, using manuals. Finally, it will be a threshold by creating new sea-scape responding to the changing environment and the speciality of workers.

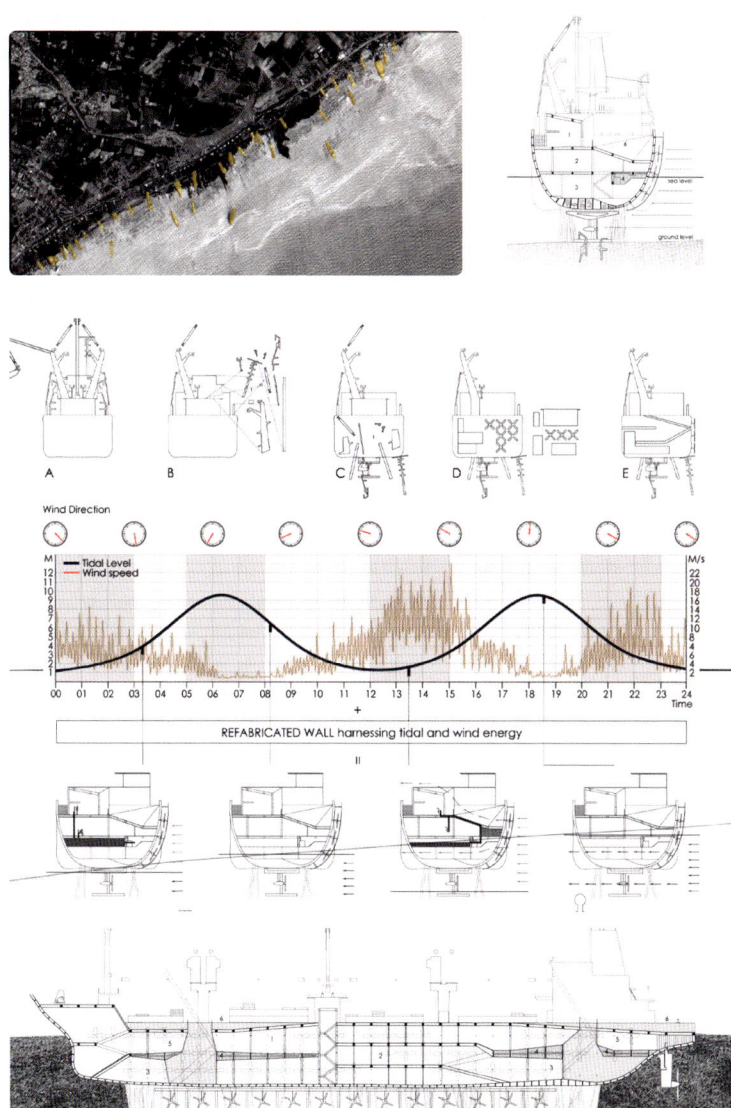

DO IT YOURSELF

AUTO- & BIOGENERATED

MATERIALS & CONSTITUTION

RECYCLED & READAPTED

MODULATED & PRE-ARRANGED

DIGITAL FAB

Wind Direction

M
12
11
10
9
8
7
6
5
4
3
2
1

— Tidal Level
— Wind speed

M/s
22
20
18
16
14
12
10
8
6
4
2

00 01 02 03 04 05 06 07 08 09 10 11 12 13 14 15 16 17 18 19 20 21 22 23 24   Time

+

REFABRICATED WALL harnessing tidal and wind energy

II

A   B   C   D   E

MORENO, David

Escuela Técnica Superior de Arquitectura de Madrid

davidmorenorubio@hotmail.com

The favelas of Rio de Janerio

The society in Rio de Janerio lives with the favela phenomenon. This kind of suburbs grows in empty territories on the outskirts of the city.

The favelas system is based on self-constructed housing. This natural growth conforms a non-planified settlement, generating a disestructured city with huge lacks of public services and facilities.

"The digitalized mountain"

The location is previously defined through a digital system. It transforms the space in function of the changing reality (topography, topology, climatology, etc.)

Each geographic point in the mountain will have a translation in a series of digital facts. Properly vectorized, these facts introduced in the software, would be able to provide the best solutions for structural and energetic systems.

"A seft-fabricated word community"

The project proposal tries to stimulate this spontaneous means of self-organization in the form of "collectives" with function as public institutions, replacing national ones and establishing their own norms to solve the current situation.

The self-construction concept extends beyond the mere physical sense. The users, by means of computer software, are able to create and design their own house.

Machines of parametric manufacture

The different parts of the housing are created by machines that produce each piece individually from a series of parametric facts. The industrialized manufacture allows the reduction of costs as long as the totally free forms for self-construction.

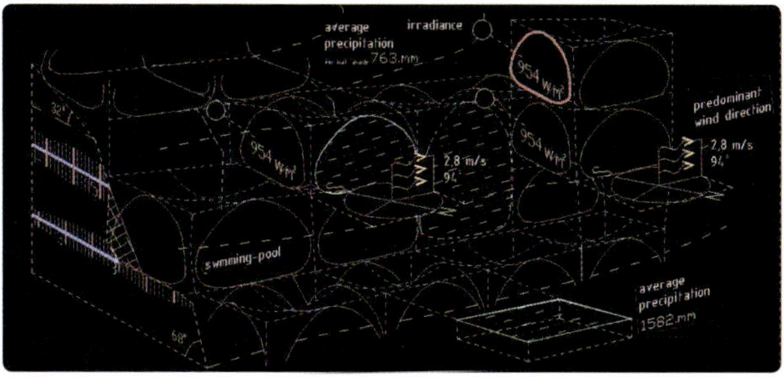

average precipitation
irradiance
763 mm
954 Wh²
954 Wh²
predominant wind direction
2,8 m/s
94°
2,8 m/s
94°
swimming-pool
average precipitation
1582 mm

solar energy
rainfall collection
natural ventilation
shading

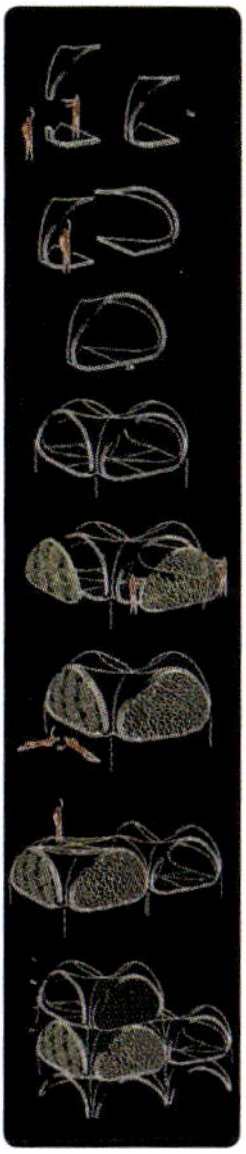

digital fab room

DO IT YOURSELF

AUTO- & BIOGENERATED

MATERIALS & CONSTITUTION

RECYCLED & READAPTED

MODULATED & PRE-ARRANGED

DIGITAL FAB

O'GRADY, Brendan
TRENOLONE, Tom

n:dL

brendan.ogrady@nocturnaldesignlab.
com

Aeroform takes advantage of a common element which is pervasive in every major metropolitan area – the billboard. It is a prototype which explores the dimensionality of the billboard and creates a subsidized living module that would allow people with limited means to live in a major metropolitan area. This fusion of the billboard and the house creates an opportunity for inhabitants to generate income from the billboard in turn subsidizing the exorbitant cost of living in the city.

Conceived as a prefabricated unit, Aeroform is constructed off-site, shipped to the site in modules, assembled, and grafted to the existing structure of a billboard. The outer shell is made of structural insulated panels covered with a layer of carbon fiber in order to reduce the weight of the structure and minimize any reinforcement to the existing billboard.

By using an existing billboard's support structure Aeroform has a minimal impact on the environment and its site. It also reduces the amount of raw materials needed to construct the home by re-using the existing structure of the billboard.

The shell, which reflects its context of the highway and the automobile, is streamlined on the exterior for optimal aerodynamic and climatic performance. Operable windows at each end of the unit allow for cross ventilation and provide an abundance of natural light within the living spaces. On the interior a programmatic shrink-wrap occurs to optimize the overall volume of enclosed space.

Since the billboard is a universal structure Aeroform can easily be adapted for use in any city, anywhere in the world.

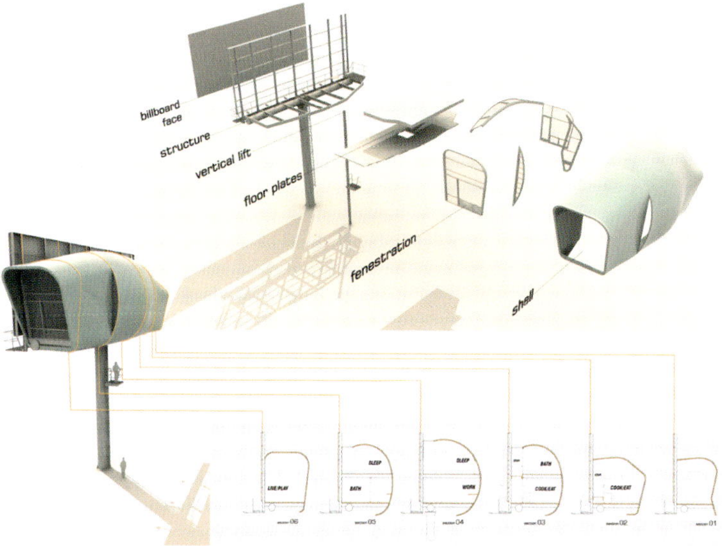

billboard face

structure

vertical lift

floor plates

fenestration

shell

USE/PLAY	SLEEP	SLEEP	BATH	BATH	
BATH	WORK	COOLEAT	COOLEAT		

section 06   section 05   section 04   section 03   section 02   section 01

DO IT YOURSELF

AUTO- & BIOGENERATED

MATERIALS & CONSTITUTION

RECYCLED & READAPTED

MODULATED & PRE-ARRANGED

DIGITAL FAB

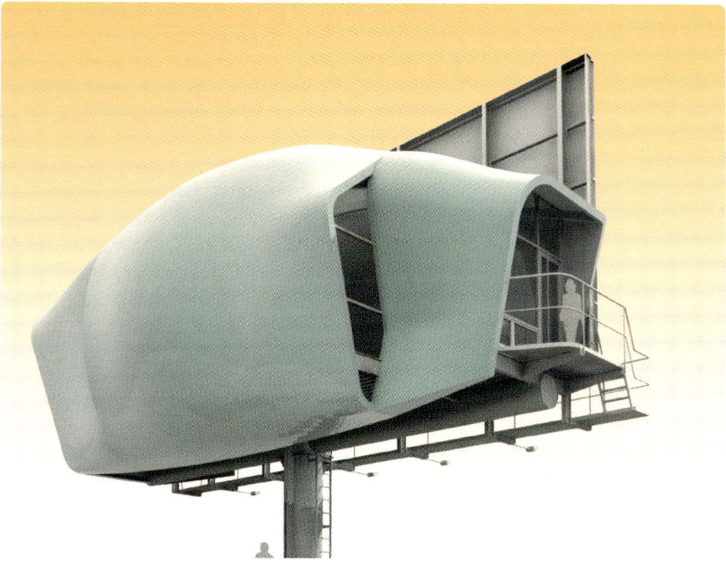

MILLER, Nathan

free agent

nmiller.arch@gmail.com

1. a collaborative website which may be directly edited by anyone with access to it.

2. a system for managing the design of a self-organized, open-source neighborhood.

Wiki Neighborhood proposes a process by which a highly specific and complex domestic network can emerge through a collaborative virtual medium. The system is enabled by the coordination of a series of concepts which are pervasive in digital culture and have the potential to redefine how communities can self-organize and self-sustain their environment. The emergence of a Wiki Neighborhood occurs at the intersection of collective interests within a virtual medium. Absolute deterministic programming is given up in favor of negotiating entangled events and situations which may then be modulated and optimized within a prototypical intelligent mesh structure.

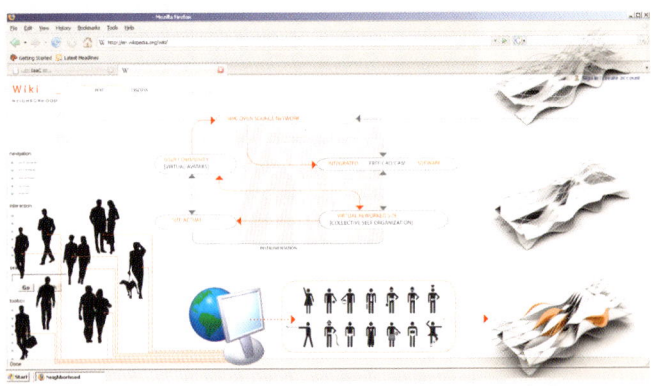

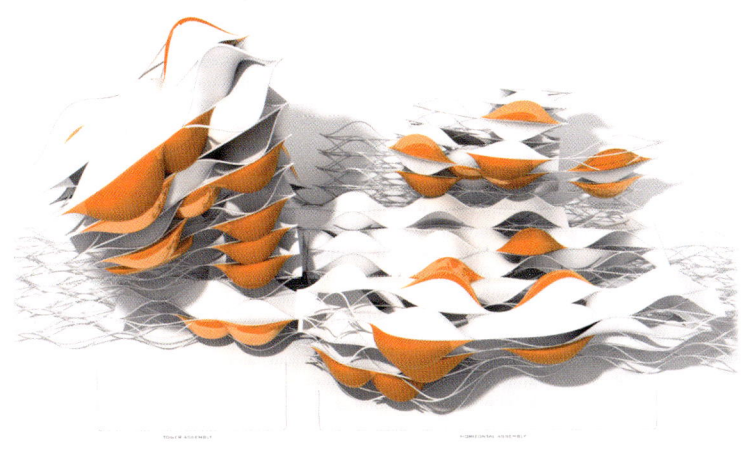

TOWER ASSEMBLY   HORIZONTAL ASSEMBLY

DO IT YOURSELF

AUTO- & BIOGENERATED

MATERIALS & CONSTITUTION

RECYCLED & READAPTED

MODULATED & PRE-ARRANGED

DIGITAL FAB

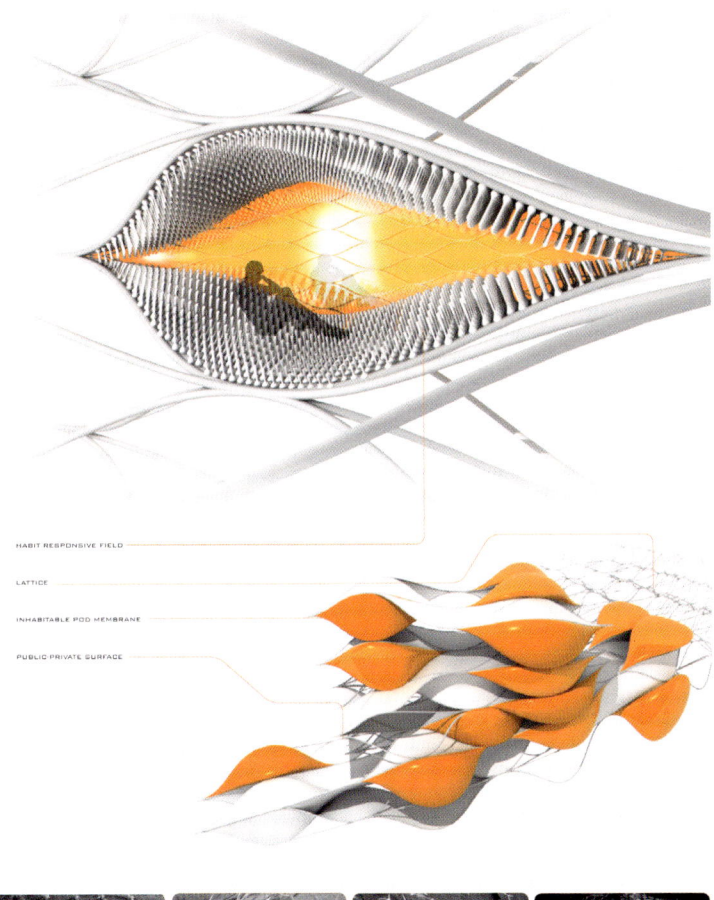

HABIT RESPONSIVE FIELD

LATTICE

INHABITABLE POD MEMBRANE

PUBLIC-PRIVATE SURFACE

DO IT YOURSELF

AUTO- & BIOGENERATED

MATERIALS & CONSTITUTION

RECYCLED & READAPTED

MODULATED & PRE-ARRANGED

DIGITAL FAB

THÀNH, V Quang   The Rain Inc

                                      vqthanh06@yahoo.com

The position of the construction is in Vietnam, the Mekong delta region, which is located in the tropical and temperate zone. It is characterized by strong monsoon influence, a considerable amount of sunny day and high-rate of rainfall, as long as humidity. There are two distinguished seasons; the hot and the rainy one.

The frame of the construction is formed from bars and joints. There are many types of bars and joints but there is only one way to combine them; between two bars there is a joint. The shell of the construction is made from series of windows that could have many axies.

To program the construction, the joint would be used as a digital component that a programable construction robot (PCR) could identify. PCR just take care of the position of the joint in space. The first node will be chosen as origin and the nodes after would be located depending on the distance from the one before. This process includes the choosing the length of bars and its orient angle. One model of construction is programmed to help PCR to choose the correct bars and joints, even the finishing panels.

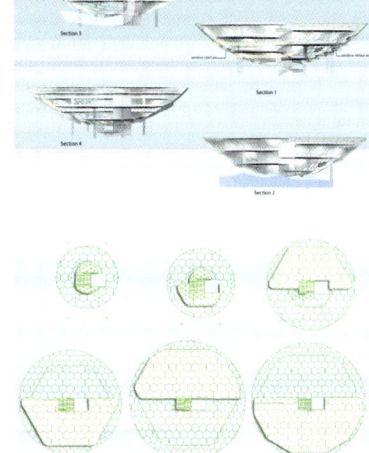

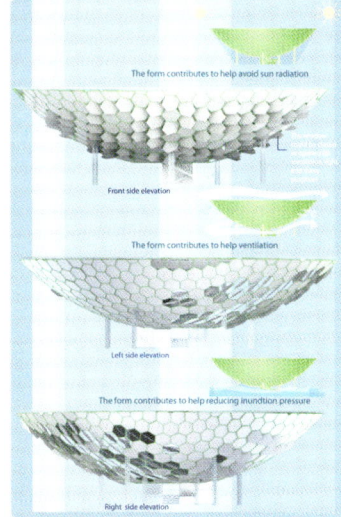

Section 3

Section 1

Section 4

Section 2

The form contributes to help avoid sun radiation

Front side elevation

The form contributes to help ventilation

Left side elevation

The form contributes to help reducing inundtion pressure

Right side elevation

DO IT YOURSELF

AUTO- & BIOGENERATED

MATERIALS & CONSTITUTION

RECYCLED & READAPTED

MODULATED & PRE-ARRANGED

DIGITAL FAB

FRESNADILLO, Javier

FREARQ

frearq@yahoo.com.ar

FACTORS: TEMPERATURE, INUNDATIONS AND LOW RESOURCES.

Theory of the chaos is the perfection between the number and the divine proportion.

The CHAOS acquires cohesion in its encounter with the territorial overlaping, ordering the randomness of the instantaneous architecture built by the man without principles of pre-configuration. Extracting their codexes, the geometry is added FRACTAL that takes charge of configuring in partial lines the structure of compound organization for its axes that manifest the flow of life in the organism in permanent construction, conceived partly of its dispersed organization, by disintegrated layers that they are extracted from the installation space. Building the hybridization among the ortogonality that contains.

The code of the place in their genetics, "expands" the scale overturning to the auto building, a final product conceived by the individual mind. Basting their threads on all the appropriation scales, the harmonic link finds polarities between two of oneself realization, the built space to human scale conceived by the individual mind and the morphological, cultural and psychosocial extractions, constitute the own morfo genesis of the collective scale that gives life to the BEING, an all organic one that possesses and grants existence to new ways of inhabiting.

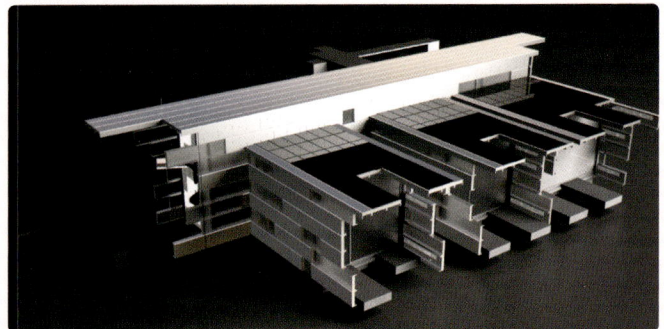

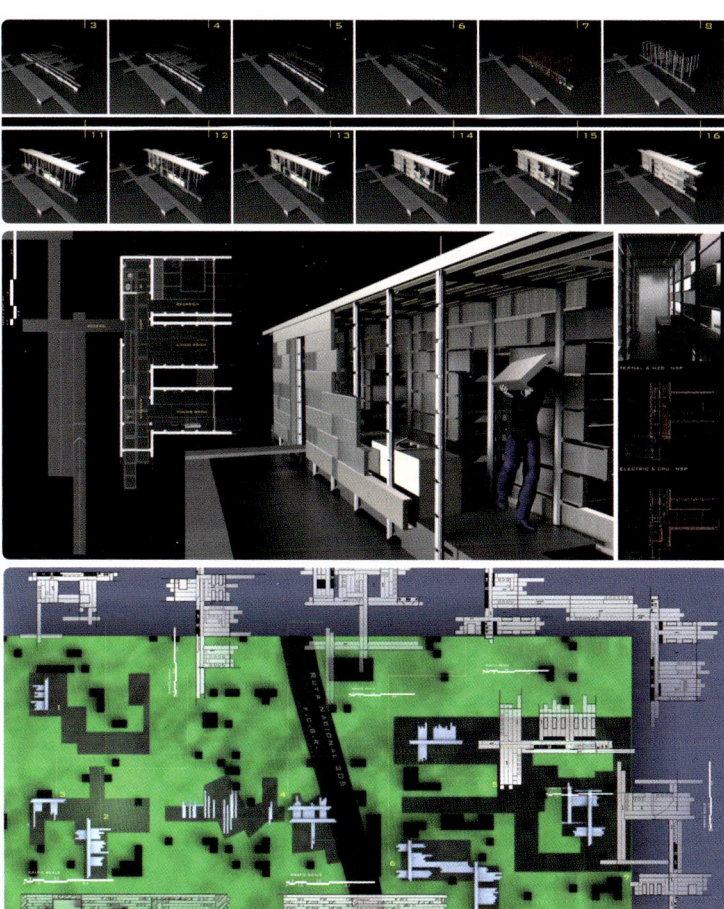

DO IT YOURSELF

AUTO- & BIOGENERATED

MATERIALS & CONSTITUTION

RECYCLED & READAPTED

MODULATED & PRE-ARRANGED

DIGITAL FAB

GARTH NG, Jason   University of Melbourne

                                       jason_0018@hotmail.com

The design of the Energy Tower and the "plug-in" modules is based on the concept of a double helix and molecules DNA is the building block of a human being, much like energy and resources is the building blocks of modern day society. A parallel can be drawn between the two, yet human ambitions for advancement has corrupted our planet and ourselves. In order to save our planet. We must look within ourselves to discover the solution.

THE CHONGQING PROTOTYPE: The first energy tower will be built in Chongqing, a city in central China with a population of 5 million. From its industrial beginnings, Chongqing is now growing into a mega city to rival Shanghai. The city is still young, but its infrastructure is aging. Now is the perfect time to build this prototype, as the start of a new age in a sustainable future. The tower has its own power, recycled water, waste management and air filtration systems. Eventually, more and more Energy Towers will be built to form a networked system of housing that will produce surplus energy and resource that will service the city, in effect creating a sustainable and near zero emission city.

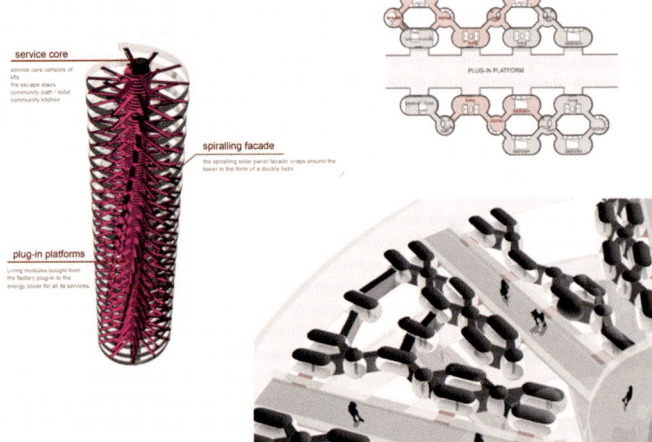

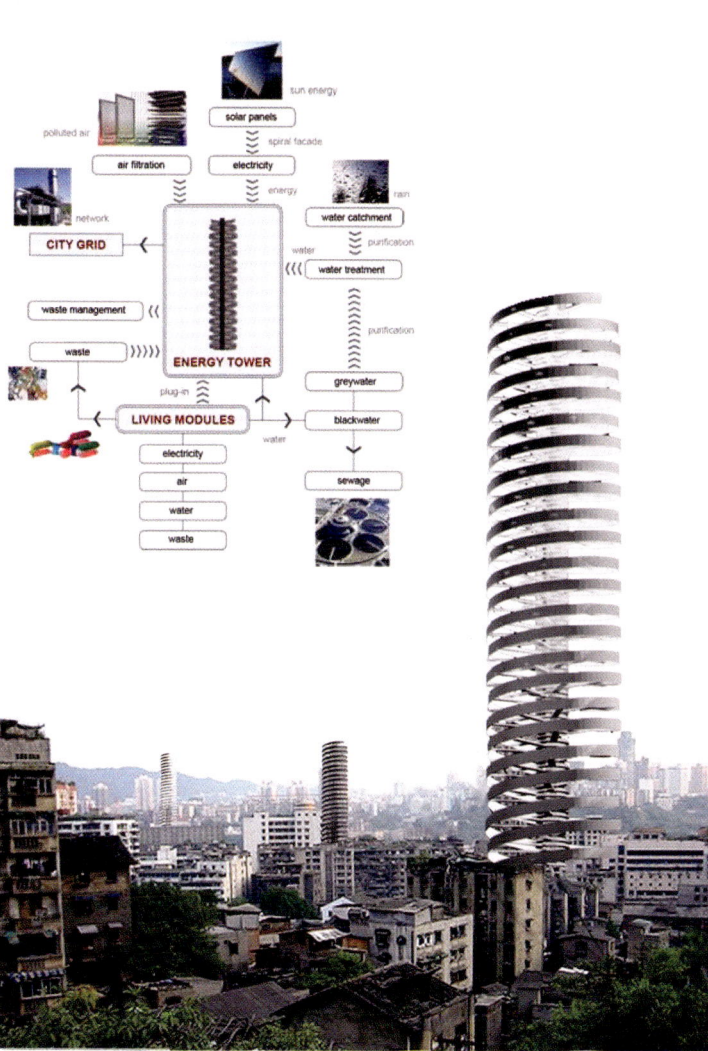

solar panels
sun energy
spiral facade

polluted air
air filtration
electricity
energy

network
CITY GRID

water catchment
rain
purification

water
water treatment

waste management

purification

waste
ENERGY TOWER
greywater

plug-in
blackwater

LIVING MODULES
water

electricity
sewage

air

water

waste

DO IT YOURSELF

AUTO- & BIOGENERATED

MATERIALS & CONSTITUTION

RECYCLED & READAPTED

MODULATED & PRE-ARRANGED

DIGITAL FAB

SMITH, Greg   P3D
KOREN, Liav
SARGENT, Michael   smith@serialconsign.com

The devastation caused by Katrina was trumped by inadequate, inefficient housing solutions. Considering the extremely unstable environmental conditions in New Orleans, a bold, forward thinking approach to housing is required. This strategy needs to deploy state-of-the-art technology in a cost-effective manner and acknowledge the inevitability of future flooding.

Instead of excavating and building a sub-structure, this housing system utilizes a series of caissons as a foundation. Hollow steel section cylinders are driven into the earth and then filled with concrete is used to repurpose fly-ash and various industrial waste.

This system is efficient and inexpensive solution to the undesirable soil conditions in New Orleans. The structural solution of this housing system relies on lightweight. The carbon fibre members are fabricated through pultrusion and shipped to the site, where they are assembled into a scaffolding system that is ready to be fitted with modular panels, concrete floorplates and mechanical systems. This system is completely modular and pieces are fabricated to include joinery and plug into the foundation.

An entire range of cladding panels has been developed that vary in transparency, density and operability. Some panel systems are able to assist in building metabolism and aid in energy collection and retention. All of these panel systems are derived through custom molds that are 3D printed with readily available agricultural by products such as starch. The expectation is that technology will soon develop to allow printing the panels outright in the near future.

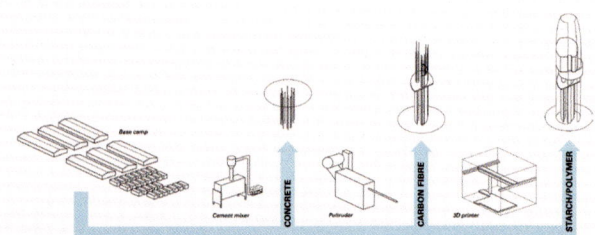

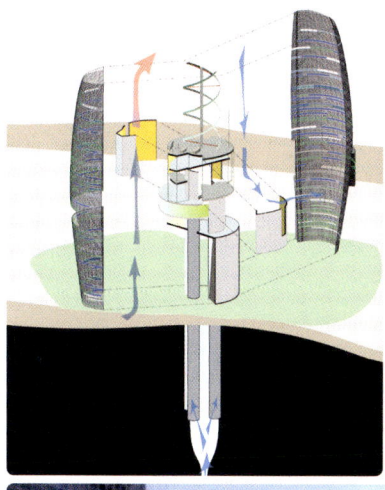

DO IT YOURSELF

AUTO- & BIOGENERATED

MATERIALS & CONSTITUTION

RECYCLED & READAPTED

MODULATED & PRE-ARRANGED

DIGITAL FAB

VICENS, Maxime
CHASSAGNOL, Antoine

Maxime Vicens Architecte

maximevicens@gmail.com

The project of "EXP architectes" aims at revisiting a type of minimum habitat, self-sufficient and self-built housing. It answers questions raised by urgency, temporary and alternative habitat, densification of urban interstices, in relation to ecological and environmental requirements.

The capsule, easily assembled by two or three people, allows a great capacity of adaptation to environment and climate: plain or mountain, desert or ice floe, urban sites, air or water. It can be winched up by helicopter if necessary. The archimedean and polyhedral morphology, a carbon tube frame holding translucent, transparent and opaque panels, permits a position adapted to the wind. The cabin can contain two residents and two extra persons, and offers sufficient volumes to eat, wash or sleep, flexible enough to allow various configurations according to the needs.

Thanks to its very limited weight and its repartition pads, no anchorage to the ground is necessary. A tank, suspended under the module, permits the treatment of 60 liters of rainwater. A wind turbine and photovoltaic panels provide the electrical supply. If assembled as a network, in the form of a fractal village, the capsules allow the energies to be shared and redistributed efficiently. They can also be used as weather stations, control or observation posts, antennas, for scientific studies etc.

> SELF-FAB PROCESS

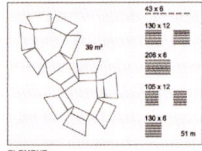

ELEMENT

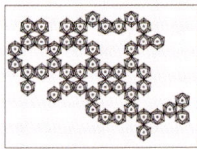

ASSEMBLAGE   FRACTAL COMMUNITIES

> SELF-SUFFICIENT SYSTEM

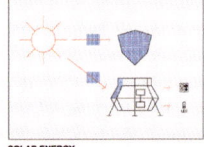

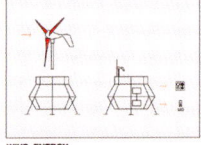

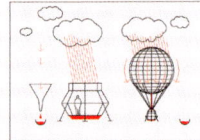

SOLAR ENERGY   WIND ENERGY   WATER RECYCLING

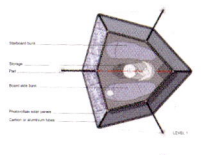

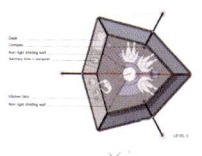

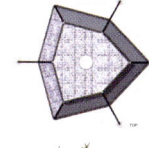

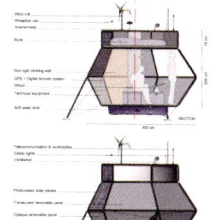

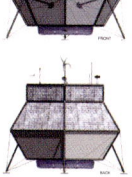

DO IT YOURSELF

AUTO- & BIOGENERATED

MATERIALS & CONSTITUTION

RECYCLED & READAPTED

MODULATED & PRE-ARRANGED

DIGITAL FAB

BLANCHI, Yann
MAGNIN, Stéphane
MALTAVERNE, Emilie
MARIN, Philippe

Research Unit
_____

philippe@research-unit.net

Our proposal is a research based on the role play both the geometry and the environmental parameters in the architectural design process. Digital fabrication make possible the realisation of non standard shapes. Thus these new kind of envelops are defined by the individual needs and by the local environmental opportunities. Moreover a networked collaborative process is proposed. This one is a support to the development of a new kind of architecture founded on innovation. Our selfabers community propose some internet online tools, guides and resources and permit to build our own habitat system. A geometric design system, an environmental design system and the use of recycled materials are proposed to the community. The online tools have the ability to produce the digital plans and the notes in function of the requirements of the user. The files are directly used by the CNC Milling or the 3D printer of the community.

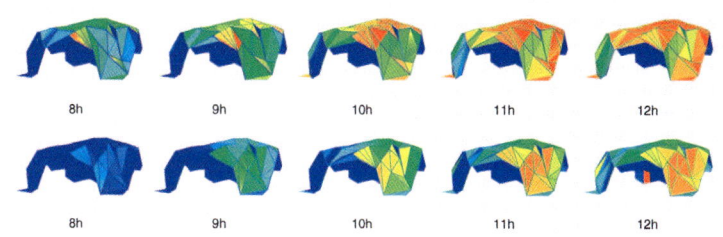

8h   9h   10h   11h   12h

8h   9h   10h   11h   12h

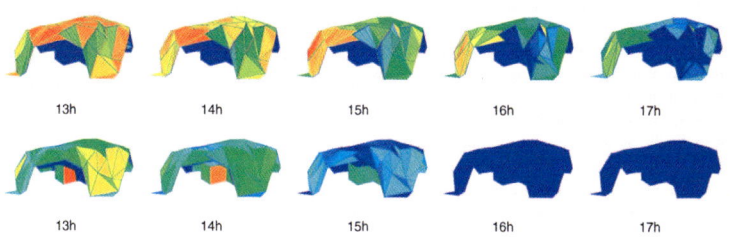

13h   14h   15h   16h   17h

13h   14h   15h   16h   17h

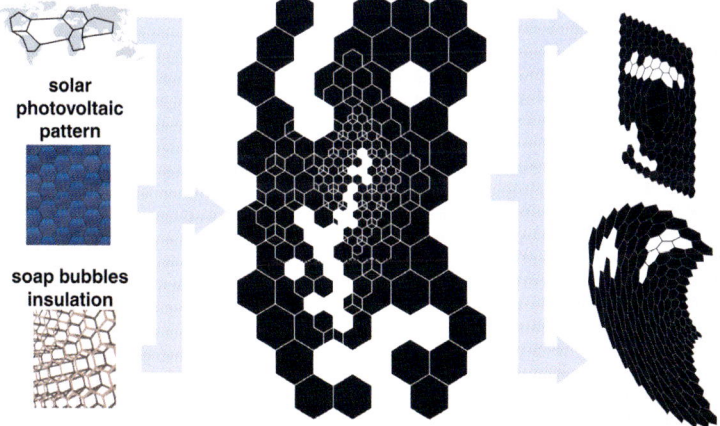

**solar
photovoltaic
pattern**

**soap bubbles
insulation**

DO IT YOURSELF

AUTO- & BIOGENERATED

MATERIALS & CONSTITUTION

RECYCLED & READAPTED

MODULATED & PRE-ARRANGED

DIGITAL FAB

QUENEY, Sebastien
CHEVANCE, Sebastien

Aion Architecture

squeney@hotmail.com

The Cocoon house offers a variable shelter to be built in symbiosis with nature: use of few recycled and natural materials for a self sufficient construction. The full process of designing and fabrication is supported by the user through specific software. The overall Cocoon is a rectangular shape that can be extended with the different needs of the user. The interior space can be modulated by stretching the middle partition. The Cocoon House is composed by a series of elements to be assembled and layered simply by one to two persons.

The main structure is composed by a series of plywood ribs laser cut by CNC machines. The number of each element is engraved to be easily assembled. Additionally, the furniture is incorporated to the structure. It consists of three layers, two are for protection from the climate and one produces energy.

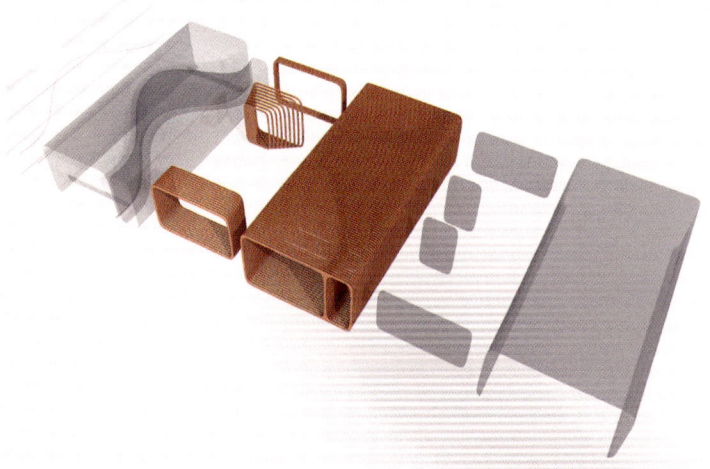

DO IT YOURSELF

AUTO- & BIOGENERATED

MATERIALS & CONSTITUTION

RECYCLED & READAPTED

MODULATED & PRE-ARRANGED

DIGITAL FAB

WILCK, Philip

University of Applied Sciences Augsburg

cutterundzirkel@yahoo.de

The Game of Life is a cellular automaton developed by the British mathematician John Horton Conway in 1970.

1. computer-aided manufacturing (CAM)

The following shown construction system is based on a computer supported design process. Those that deliver concrete parametric data in approach to mechanical manufacturing. An economic way to produce this specific form, with all the variant parts, will be possible with the computer-aided manufacturing method.

The supporting structure consists of chamfered and bended binder, these will be produced with industrial common CNC cutter. This method is well known from the engine construction and has found its way little by little into architecture. It is still a dream that everybody will be able to, by submitting the essential parameter, create his own home with these modern equipment.

2. environmental relations

Using the earth as a heat source/ sink, a series of pipes, commonly called a "loop", is buried in the ground near the building to be conditioned. There are many ecologic and economic ways to supply the inhabitants of the "Game of life" house. Water, energy and food can be produced in the "Game of Life", this aids strongly to discharge the environment.

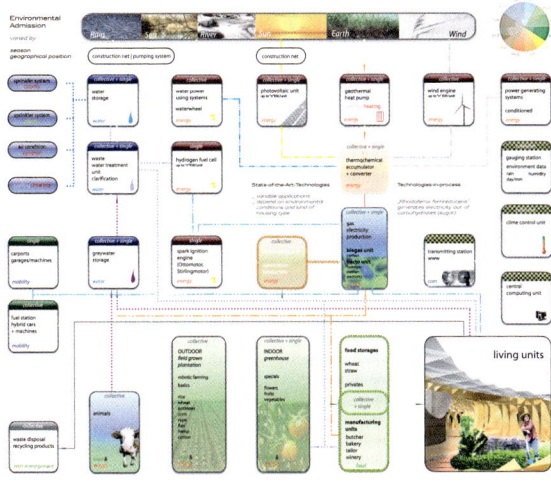

living units

DO IT YOURSELF

AUTO- & BIOGENERATED

MATERIALS & CONSTITUTION

RECYCLED & READAPTED

MODULATED & PRE-ARRANGED

DIGITAL FAB

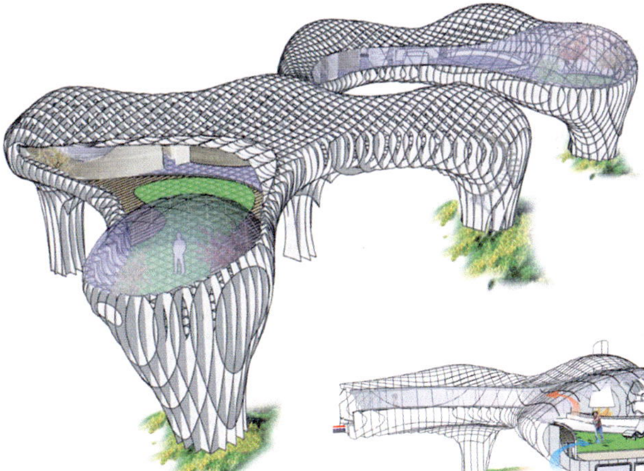

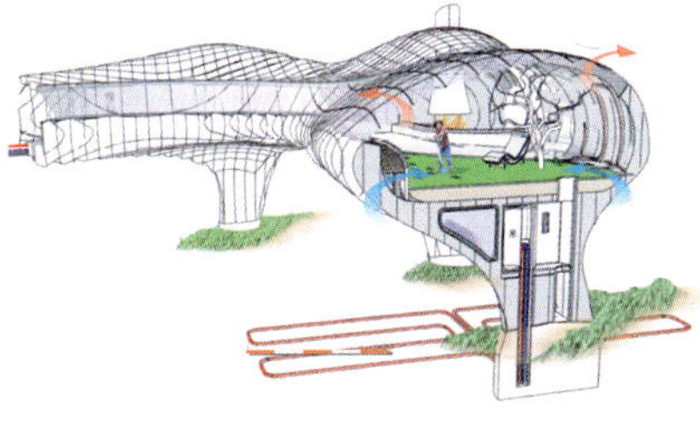

GOURDOUKIS, Dimitris   object-e architecture
TRYFONIDOU, Katerina

                                  object.e.architecture@gmail.com

The process begins with the generation of random elevation patterns. Those patterns are getting tested in a light simulator software (ecotect) for different locations, seasons and times so an initial database of solutions to be created. Each time that units have to be produced, random patterns are tested until a satisfactory solution is found. The tests are added to the database. The feedback of the users is used in order to re-evaluate the data collected.

The production of new units is becoming a self learning process for the program. At the same time the users with their feedback are altering the initial standards in a never-ending, multiple exchange and altering of information. The growth of the database is strictly related to the growth of the inflate it settlements and follows similar self-referential patterns. The data stored in the database gets transformed from data regarding lighting strategies to data regarding relationships

— flexible metal structure

— rubber

— recycled fabric. provides shadow

— inflatable membrane.

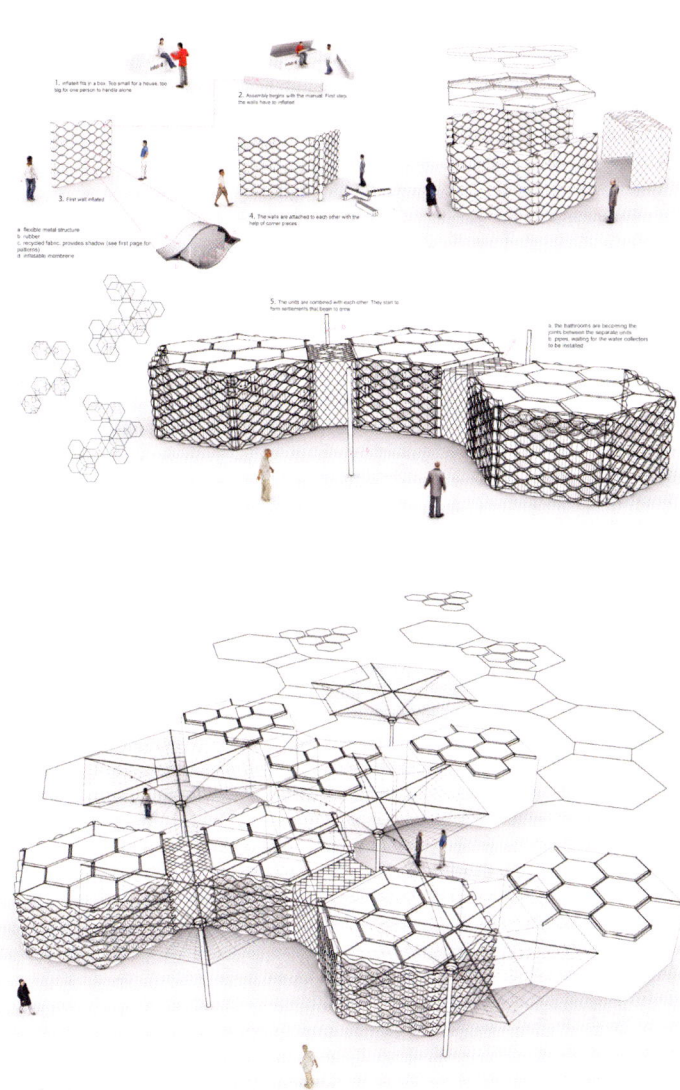

1. Inflatant fits in a box. Too small for a house, too big for one person to handle alone.

2. Assembly begins with the manual. First step, the walls have to inflated

3. First wall inflated

4. The walls are attached to each other with the help of corner pieces.

a. flexible metal structure
b. rubber
c. recycled fabric, provides shadow (see first page for patterns)
d. inflatable membrane

5. The units are combined with each other. They tend to form settlements that begin to zone.

a. the bathrooms are becoming the joints between the separate units
b. pipes, waiting for the water collectors to be installed

DO IT YOURSELF

AUTO- & BIOGENERATED

MATERIALS & CONSTITUTION

RECYCLED & READAPTED

MODULATED & PRE-ARRANGED

DIGITAL FAB

IaaC by way of four combats:
visions in the framework of advanced architecture.

Winners of the first Advanced Architecture Contest, currently researching at IaaC, Daniel Ibañez and Rodrigo Rubio.

COMBAT 1
'MULTI' vs. DISCIPLINARY

We are many, from many places and with many specialties.

We start from the many and the various, believing in efficiency, intelligence and the emergence of non-hierarchical networks. Believing that an approach from difference and the multiple will get us closer to the critical and the necessary.

The multiscalar approach to the project for the redefining of reality is one of the most fundamental principles of our way of working. A strategic global approach to the problem can identify the most crucial areas of opportunity. We work simultaneously from the scale of fabrication to the territorial scale, continually moving from one to the other, as a mechanism for transforming the discipline.

We also believe in the pooling of interests, in negotiation between individuals with opposing interests, and we read the IaaC as an environment for international participation, an assemblage of institutions, companies, professionals and students who approach their work from multicultural and multidisciplinary perspectives and address issues that are both global and local. An environment in which engineers, anthropologists, sociologists, mathematicians, economists, biologists and ecologists are indispensable allies in generating the necessary emergence of new ways of understanding and constructing reality.

We thus construct our position as the product of a dispersed deployment.

Different nodes, specialized in different disciplines and working together arrive at a more robust and multifaceted position, developing knowledge bases in multiple directions.

FAB LAB vs. GLOBALIZED INDUSTRIALIZATION

Materials from here plus global knowledge.

We see the new paradigm shift as our opportunity for action. In the unresolved conflict between the local and the global. We respond to the contemporary need for individualization and adaptability. Assembly-line techniques, mass production and linear processes are of no use to us now.

And we believe in action. In trial-error loops. In distributed participation. And so, thanks to an initiative of the MIT and its Center for Bits and Atoms, we have a global network of digital fabrication laboratories (Fab Labs).

This gives society at large access to the most advanced technologies. The black box has been opened, effectively integrating the user into the processes of construction. Prototyping techniques are now readily available and becoming a completely open mode of social expression instead of the exclusive preserve of certain technically privileged sectors.

Each Fab Lab is thus a small workshop, but with the infrastructure and tools needed to produce almost anything. Each one located somewhere in the world, each one specializing in a scale of production ('from bits to territories'), and all of them working in a network using the same protocols and languages. The result is a heterogeneous but coherent network.

While they cannot compete with large-scale production and distribution, they offer a freedom and creativity that big production systems and their business interests do not permit. In this way they are able to adapt to the most local needs and individual intentions, resulting in more advance solutions.

OPEN SOURCE vs. COPYRIGHT

The value is not in the rights.

We believe, then, that design, like the redescription of reality, must be an open process, and far from sectorized or technocratic concepts, we are committed to everyone having a voice, seeing the result as the resolution of negotiations and continuing additive modifications.

We would like to borrow the concept of open source used in software design. We would like to transplant it to our discipline, sharing our activity with partners and users. The state of the object/project is never a closed end point but a frozen diagram of the state of negotiations between priorities of energy, of materials, industrial priorities, domestic priorities...

The transparency of process that is essential in open source projects also makes for a fuller understanding of the process. And a deeper understanding always leads to deeper engagement, social involvement and social awareness, participation... Activation and multiplication of the network. What is important is the generation and transmission of shared knowledge.

SELF-FABRICATION vs. MASS-CONSTRUCTION

Do it yourself, just as you want it. And to achieve all this we need, once again, to rethink the scales. While large scales offer strategic values, reducing the scale produces adaptability, faster responses. We propose to effect a change in the scale of construction systems and processes. From the scale of the crane to the scale of the man. By reducing miscellaneous equipment we control the consumption of energy and materials and have more flexibility and room for manoeuvre. The use of non-standard materials and techniques, which are closer to raw materials and craft processes, allows us to give a more precise response. Greater adaptability.

Once again, it allows us to involve the users in the process of configuring their most immediate habitat, thereby facilitating the understanding of the architectural object.

For this reason, the use of personal fabrication techniques, lightweight construction systems, adaptable and ecological materials, etc... is important not only for environmental reasons but also for the understanding that these processes generate in their social niche.

Sustainability thus passes from the technical to the domestic sphere.

ID code · Country · *Name*

01e8b9 · Slovakia · *Tothova, Sona*
025678 · Netherlands · *van Lidth de Jeude, Marije*
035ba7 · VietNam · *Hoang, Minh*
03af0c · KoreaRepublic · *Song, Kwanghyun*
03f1f4 · Brazil · *Luz Vasconcelos, Vanessa*
0756a7 · Spain · *Vecino, Victor*
079cea · Mexico · *Mendoza Cruz, Gabriel*
08ca58 · India · *Ektate, Sanjay*
0acffc · Portugal · *Duarte Bento, Pedro*
0b05e0 · UnitedStates · *O'Grady, Brendan*
0bd50b · Italy · *Donner, Luca*
0d70b9 · Colombia · *Londoño Ramirez, Adrian Alberto*
0dc65b · Romania · *Barsan Pipu, Claudiu*
0e3beb · UnitedStates · *Bross, Zan*
0e7e61 · Iran · *Mozaffari Kakavandi, Arash*
0ee28b · Colombia · *Figueroa, Juan*
0f9b63 · Spain · *Rodriguez, Xavier*
0fbcfc · India · *N siddharth*
13bb21 · VietNam · *V. Quang, Thành*
166dd9 · Finland · *Gelard, Philippe*
174e92 · France · *Hinfray, Victoire*
1788ff · Israel · *Almani, Moshe*
175fde · Australia · *Ko, Gary*
177210 · HongKong · *Chi Ho, Fung*
18f0cc · Algeria · *Mezoued, Aniss*
1c97cc · Mexico · *Gonzalez Rene, Erik*

1d794c · Argentina · *Vezzosi, Guillermo*
211308 · Indonesia · *Yoas Sihotang, Jonathan*
2202c1 · UnitedStates · *Pierce, Jason*
226437 · Spain · *Pineiro Garcia, Ana*
226a76 · India · *Kanth, Chandra*
230d5b · SerbiaMontenegro · *Gavrilovic, Marko*
23852d · FYROM · *Pazardzievska, Elena*
24e419 · Sri Lanka· *Suriyaarachchi, Tharinda*
27bfed · Mexico · *Valencia, David*
283142 · Denmark · *Aagreen, Line*
28f24d · France · *Routier, Adélaïde*
2c10ef · Thailand · *Yamkleeb, Jirawit*
2ce5d1 · Spain · *Torre Sanchez, Roman*
313473 · Spain · *San Gregorio, Sara*
346a1a · Malaysia · *Chee Sheng, Ang*
36ebac · Mexico · *Orozco, Melisa*
377017 · Australia · *NG Garth, Jason*
38cdeo · Spain · *Diaz Tarrisse, Daniel*
3ac77a · France · *Mansuy, Corina*
3c8d5f · Germany · *Jessnitz, Christoph*
3ccd8a · China · *Cao, Yuan*
3cd9eo · Mexico · *Ponce Loya, Enrique*
3cfe56 · Malaysia · *Wong Yoke Fong, Doreen*
3f1f37 · China · *Hailong, Wu*
416ddc · Mexico · *Domenzain, Carlos*
41d697 · China · *Lei, Zhou*

5c64bb · Brazil · *Silva, Eliane*

5dec56 · Spain · *Garcia Martinez, Angel*

5e3768 · UnitedStates · *Marcum, Charles*

5e5213 · Indonesia · *Adisaputra, Risanda*

5f7266 · Spain · *Moreno, David*

5f88bc · Thailand · *Jittakasem, Suebsai*

60564e · Chile · *Muñoz Bahamonde, Pablo Javier*

607873 · China · *Huang, Jun*

61aff3 · Bangladesh · *Khan, Nabi*

62da37 · India · *Kannan, Mathan*

63c3c2 · France · *Marin, Philippe*

643284 · Germany · *Gruss, Hendrik*

662169 · Brazil · *Barbi, Bruno*

6765bc · Italy · *Emili, Ana Rita*

679d49 · Colombia · *Seba, David*

67f8ee · Colombia · *Chamie Garcia, David*

69b97d · Japan · *Yoshizawa, Ikuma*

69c1ac · Poland · *Melion, Aleksandra*

6a4775 · Brazil · *Andrade, Danilo*

6a570b · Brazil · *Santos, Esdras*

6b4315 · Spain · *Berasategui Canals, Marina*

6c66d4 · United States · *Teigan, Annsaint*

6d2323 · UnitedKingdom · *Kerimol, Levent*

6e74cb · Colombia · *Morales, Ivan*

6fe322 · UnitedStates · *Taylor, Cory*

6fefd9 · SerbiaMontenegro · *Cukic, Iva*

70c818 · Lebanon · *Tarek, Salloum*
70b118 · Colombia · *Guevara, Eliana*
7161c5 · Mexico · *Larios Sanz, Rodrigo*
72014f · KoreaRepublic · *Nam, Gun Wook*
728fdf · Mexico · *Alvarez, Fernando*
73320e · Chile · *Diaz Cadiz, Yerko Andres*
73bd50 · Mexico · *Saldivar, Alfredo*
740a84 · Russian · *Kozlov, Roman*
7514e6 · Mexico · *Jacobson, Olaf*
760183 · PuertoRico · *Arroyo Pagan, Edgardo*
76c32f · Germany · *Wilck, Philip*
77c376 · CostaRica · *Zamora, Marco*
78230b · Mexico · *Barba Castillo, Fernando*
78daa8 · Mexico · *Flores Villalobos, Ivan*
78fd0d · Spain · *Rivas Ruzafa, Elena*
791da7 · Mexico · *Pantoja, Rodrigo*
7d3eed · Italy · *Benato, Davide*
7db689 · Bangladesh · *Sabbir, Ahmed*
7dc4bc · Turkey · *Yilmaz, Emre*
7e00f0 · Italy · *Raggi de Marini, Fiammetta*
7f776c · UnitedStates · *Caldwell, Kyle*
7f78af · Mexico · *Pena, Adriana*
8255ad · HongKong · *Ho, Kenneth Kai Kit*
857ab2 · Mexico · *Pacheco, Roberto*
85c655 · UnitedStates · *Vann, Andy*

86a4c0 · Peru · *Boljsakov, Natalija*
86fa95 · Mexico · *Madrigal, Maui*
87575f · India · *Rajendran, Rajnirmal*
87fe69 · Mexico · *Contreras Loreto, Roberto*
888934 · Mexico · *Amparan Infante, Armando*
88da11 · Brazil · *Altera Oliveira, Maycon*
8912a3 · Turkey · *Ünal, Pinar*
8971b1 · France · *Queney, Sebastien*
89b00b · Italy · *Bellardi, Francesco*
8aca97 · Egypt · *Gerisha, Hesham*
8bf42b · Mexico · *Gallegos, Denis*
8ca351 · Mexico · *Ortega Stark, Alejandro*
8e3365 · UnitedStates · *Marquez, Sergio*
8eb949 · France · *Delchet, Aurelien*
8ee5d4 · China · *Gu, Hao Qi*
8f5c54 · France · *Vicens, Maxime*
8f97cd · SerbiaMontenegro · *Draskovic, Hana*
8fc967 · United States · *Sibille, Nicholas*
9057c9 · Colombia · *Saldarriaga, Juan*
906aa7 · Poland · *Dadok, Paweł*
9089ac · China · *Xin, Li*
90995f · Argentina · *Forciniti, Federico*
9128d3 · Brazil · *Soares Garcia, Luciana*
92926b · SerbiaMontenegro · *Damjanov, Nikola*
92c868 · Mexico · *Sepulveda, Ruben Octavio*

9342cc · France · *Alchie, Laure*
93441b · Germany · *Rex, Manuel*
934802 · Algeria · *Kacem, Rai*
936432 · Egypt · *Hafez, Saleh*
9931fe · Peru · *Bartesaghi Koc, Carlos*
99a9d9 · United States · *Hromada, Jan*
99de2f · Mexico · *Alvarez, Jose*
9a7bb9 · Germany · *Henrich, Elena*
9ab73c · Philippines · *Caumeron, Jimmy*
9b4183 · Mexico · *Anton, Juan Carlos*
9bc4a9 · Australia · *Welopo, Handri*
9f17f6 · Italy · *Pagnini, Lorenzo*
9fb492 · France · *Rochat, Alberto*
a0e994 · UnitedStates · *Moore, Ashley*
a11832 · Israel · *Cory, Joseph*
a131c7 · KoreaRepublic · *Lee, Junshick*
a1f723 · Italy · *Liotta, Salvator-John*
a376fe · Netherlands · *Langenberg, Ernü*
a40c95 · Netherlands · *van Dijk, Marcel*
a836a3 · Sweden · *Rubing, Anders*
a90365 · Mexico · *Gomez, Carlos*
aa081f · Spain · *Cabanas Ballbè, Miriam*
aa81a4 · CostaRica · *Schuette, Oliver*
ab7903 · Italy · *Vartellini, Oreste*
ad4914 · Mexico · *Juarez, Alberto*

aead89 · Spain · *Zaragoza Cuffí, Marcos*

af5b55 · KoreaRepublic · *Lee, Seung Teak*

b08827 · Spain · *Ballesteros Simón, Ivan*

b0adec · Mexico · *Ortiz Uribe, Jorge*

b283f0 · Spain · *Hernandez Fernandez, Ruben*

b3410e · UnitedStates · *Deans, Scott*

b43e15 · Argentina · *Arjol Acebal, Ignacio*

b68782 · Brazil ·*Umbelino Ferreira, Jaqueline*

b6c9ea · Greece · *Olympios, Kyriakos*

b71c56 · Egypt · *Abu Elso'oud, Yaser*

b82e1b · United States · *Keerns, Ryan*

b8c1d1 · Brazil · *Pedro, Vanessa*

b9b92b · Uruguay · *Rodriguez Sanchez, Claudio*

bdfb78 · Slovakia · *Malaga, Peter*

be5ca3 · United States · *Speelman, Alexander*

bf268d · UnitedStates · *Bergeron, Gabriel*

bfe4ee · Japan · *Matsuoka, Satoshi*

c2b33c · Lithuania · *Jonauskis, Tadas*

c2bd4e · HongKong · *Okuda, Shinya*

c2e0cb · Spain · *Gil Moreno de Mora Macián, Patricia*

c30fb2 · UnitedStates · *Torres, Luis*

c32e36 · France · *Bertina, Stéphanie*

c36d77 · Singapore · *Wirawan, Gusti*

c430c2 · Brazil · *Nascimento, Lúcia*

c42c4b · Canada · *Salmon, Jeff*

e152b2 · CostaRica · *Liu, Henry*
e17488 · SerbiaMontenegro · *Djermanovic, Predrag*
e2db6e · Italy · *Ulisse, Alberto*
e5b3fb · Brazil · *Gonçalves, Keila*
e60100 · Germany · *Schuette, Oliver*
e62a89 · Germany · *Klein, Tobias*
e781dc · Taiwan · *Kuan Yu, Lin*
e9c45f · Turkey · *Tavli, Deniz*
e9c5aa · SouthAfrica · *Hatzifotiadis, Anastasio*
eabea7 · France · *Bedu, Olivier*
ec3532 · Spain · *María, Antón*
edf74d · Canada · *Yeung, Wai Yip*
edfe6a · Spain · *González, Airam Eloende*
f1c243 · China · *Tang, Ming*
f26661 · Japan · *Shinichiro, Takahashi*
f35501 · Argentina · *Yenerich, Adán*
f409b0 · Lithuania · *Stankevic, Jezi*
f75c80 · Greece · *Katsikis, Nikolaos*
f79b89 · VietNam · *Le Xuan, Bach*
f8fd15 · China · *Wenjun, Li*
f9b4a9 · Canada · *Smith, Greg*
fcbe42 · Turkey · *Çakır, Onurcan*
fddf49 · Philippines · *Taclob, Myrtle*
fee4a0 · Italy · *Stevan, Luisa*
ffcbaf · France · *Charrat, Chrystel*

Faculty Members

Vicente Guallart
Willy Müller
Marta Malé-Alemany
Izaskun Chinchilla
Aaron Betsky
Lucas Cappelli
Jose Pedro Sousa
Robert Brufau
Andreu Ulied
Uriel Fogue
Neil Leach
Artur Serra
Vincent Julian
Olaf Gipser
Martin Sobota
Lluis Viu Rebés
Andrés Jaque
Gonzalo Delacamara
Ferran Grau
Michel Rojkind
Bostian Vuga

Florian Foerster
Sabine Müller
Andreas Quednau
Max Sanjulian
Victor Viña
Gerard Passola
Jordi Pages i Ramon
Jorge Aleix
Daniel Ibáñez
Rodrigo Rubio
Shane Salisbury
Luis Fraguada
Areti Markopoulou
Francisca Aroso
Nacho López
Toni Moranto
Orfeas Giannakidis
Moises Morato
Spyros Stavoravdis
Joan Miralles

2nd Advanced Architecture Contest

Director
Lucas Cappelli

Coordinator
Luciana Asinari

Contents Advisers
Marta Male-Alemany
Willy Müller
Jennifer Mack
Isabel Castro Olañeta
Gaston Jorge Gaye
Turlif Vilbrandt
Orfeas Giannakidis

Collaborators
Asaduzzaman Rassel
Areti Markopoulou
Florise Pages
Silvia Brandi
Chris Kemper
Jesus Lara
Hongao Zhao
Chiara Farinea
Anna Szloser
Abel Patacho
Monica Tadeo
Martha Mazzoni

Communication & Web
nitropix.com
Monica Tadeo
Jorge Ledesma
Mateo Lima Valente

Jury Members
Yung Ho Chang
Turlif Vilbrandt
Young Joon Kim
Michel Rojkind
Josep Lluís Mateo
J.M. Lin
Julio Gaeta
Greg Lynn
Vicente Guallart
Lucas Cappelli
Willy Müller
Marta Malé-Alemany
Rodrigo Rubio
Daniel Ibañez

Editors
IaaC
Institute for Advanced
Architecture of Catalonia
Actar

Responsibles for the Edition
Lucas Cappelli
Vicente Guallart

Coordination
Tomas Diez
Vagia Pantou

Text Classification
Vagia Pantou

Translation
Graham A. Thomson

Graphic Design
Christian Schärmer
for ActarPro

Production
Actar Pro

Printing
Ingoprint S.A.

Distribution
Actar D
Roca i Batlle 2
08023 Barcelona
T +34.93 41 74 993
F +34.93 41 86 707

Collaborators

 MINISTERIO DE VIVIENDA

 fundació caixa d'arquitectes

Riera Urbanizer

ISBN 978-84-98954-74-8
DL B-27965-2009

Printed and bound in the European Union

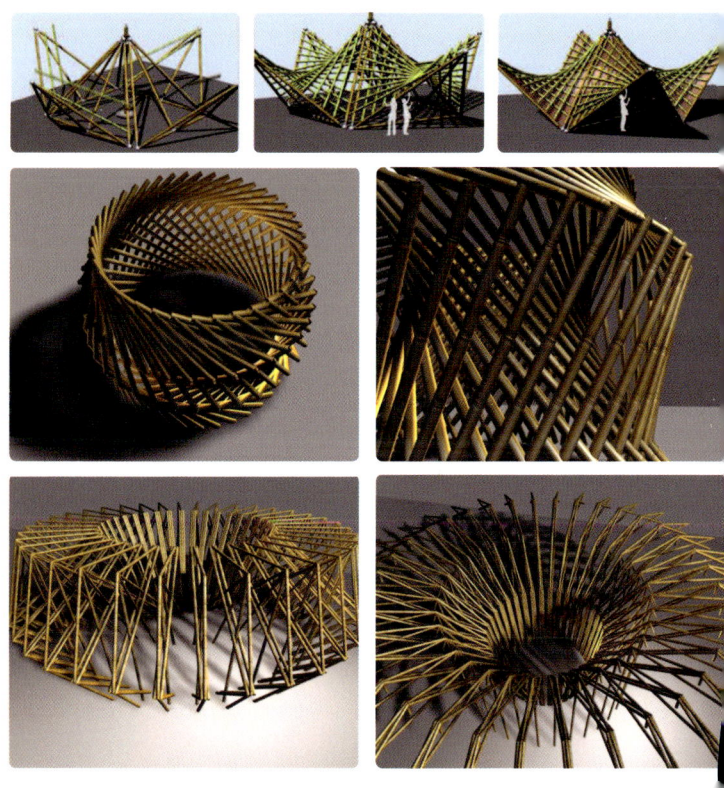

The Self-Fab House is the result of the 2nd Advanced Architecture Contest organized by the Institute for Advanced Architecture of Catalonia, IaaC. It presents more than 100 projects for new ideas and solutions of making houses in a more efficient way, using cleaner technologies, and reducing the impact of the construction process on the global environment, organized by 6 categories, including Do It Yourself, Auto & Bio Generated, Materials & Constitution, Recycled & Readapted, Modulated & Prearranged, Digitally Fabricated.

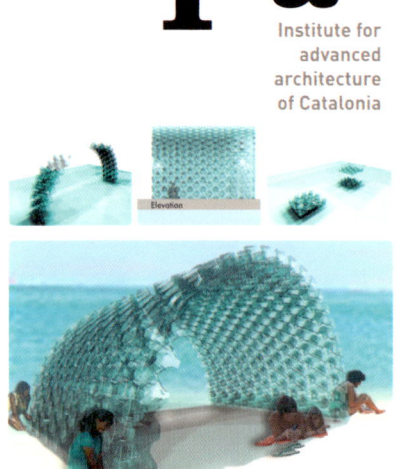

Institute for
advanced
architecture
of Catalonia

ISBN: 978-84-96954-74-8

UNITED NATIONS INDUSTRIAL DEVELOPMENT ORGANIZATION

INTERNATIONAL CENTRE FOR SCIENCE AND HIGH TECHNOLOGY

**Proceedings of the Second International Course on
Research and Innovation Management**

Manual

*for Training in Research and
Innovation Management*

Moscow, Russia   29 October – 8 November 1992

Editor

Augusto Forti

International Centre for Science and
High Technology, Italy

World Scientific
Singapore • New Jersey • London • Hong Kong

Published by

World Scientific Publishing Co. Pte. Ltd.

P O Box 128, Farrer Road, Singapore 9128

USA office: Suite 1B, 1060 Main Street, River Edge, NJ 07661

UK office: 73 Lynton Mead, Totteridge, London N20 8DH

MANUAL FOR TRAINING IN RESEARCH AND INNOVATION MANAGEMENT
Proceedings of the Second International Course on Research and Innovation Management 1992

ISBN 981-02-1653-X

Printed in Singapore by Utopia Press.

Introduction

Science and technology are among the key driving forces of change and development of today's societies. Enormous investments in scientific research, technology and innovation are made by governments and private industries. An important new characteristic of this development is the increasing scientification and knowledge content of industrial products and processes. Scientific research in universities and independent institutions has moved closer to industry in order to gain in efficiency of commercialization. The more society invests of its financial and human resources in scientific and technological research, the stronger will be the demands by society and government for quick returns on investments. This requires more effective management of research and innovation. An additional dimension is the dramatically increasing international competition. It seems that Japan, better than other countries so far, has succeeded in the effective transformation of the pool of scientific and technological knowledge into industrial products for a world market.

To address these issues and prepare scientists to tackle successfully the problems connected with the management of modern research activities, the International Centre for Science and High Technology (UNIDO-ICS) is organizing an annual International Course on Research and Innovation Management.

The institutes involved in this course are the following:

International Centre for Science and High Technology (ICS), Trieste, Italy; UNESCO Regional Office for Science and Technology for Europe (ROSTE), Venice, Italy; European Institute for East-West Co-operation (EIEWC), Venice, Italy; London School of Economics and Political Science (LSE), London, United Kingdom; Innovation Institute, Stockholm, Sweden; Commission of the European Communities (CEC), Brussels, Belgium; University of Trieste, Trieste, Italy; Italian National Research Council (CNR), Institute for Studies on Scientific Research and Documentation, Rome, Italy; Russian Academy of Sciences (RAS), Moscow, Russia; Moscow International Energy Club (MIEC), Moscow, Russia; and other National and International Institutions.

The Course is tailored following the needs of the country in which it is going to take place, taking into account the cultural and social tradition of the country itself.

To date we have offered two modules, one limited to 10 days of intensive course work, and the other lasting 20 days.
The main subjects of the Training Course are the following:

- Managing Change and Uncertainty: How Information Systems Can Help
- IT Tools for Project/Laboratory Management
- Project Management
- Technical and Economic Monitoring of R&D Projects
- Marketing of Scientific Products
- The Environment of Research, Development & Innovation
- Creativity and the Innovation Process
- Financing of Research, Innovation and New Product Development
- Strategic Management of Research and Innovation.

The Course is designed for directors of research laboratories, leaders of research projects, R&D Managers from private or public sectors, policy makers in the scientific and technological field and there is a very strong international demand, particularly from developing countries, for this type of training, to which the international community should be more responsive.

The Course was received with great interest from Russian participants coming from public and private enterprises in Russia. In particular, the Ministry for Science, Higher Education and Technological Policy has asked ICS to assist in the establishment of a Centre for training and research in the innovation process.

Augusto Forti
Moscow, December 1992

Table of Contents

The material, compiled by Ms. Lioudmila Kouzminova, is an abstract of the Proceedings of the *Second International Course on Research and Innovation Management* (29 October – 8 November 1992, Moscow, Russia).

Living with Uncertainty, and Loving it

Ian O. Angell, Department of Information Systems, London School of Economics and Political Science, London, United Kingdom

Change, Uncertainty and Tidiness

Today's businessmen are facing a contagion of profound uncertainty. Their knee-jerk reaction is to insist that the world is tidy. They have a *"desire to deprive the world of its disturbing and enigmatic character."* The management literature is full of cure-alls that sell tidiness to the gullible manager in the form of ritual application of systems and methods. This mind-set of constantly tidying up, a recognized psychological disorder, called 'obsessive compulsive neurosis', is increasingly apparent as a corporate disposition. *"Madness is something rare in individuals – but in groups, parties, peoples, ages it is the rule."*

This neurotic call for tidiness has been seized upon as a marketing opportunity by consultants and systems analysts. They charge onto the scene, promising a world of functionality, neatly described in networks of boxes, triangles, hexagons, circles, and lines; a world controlled by 'bubbleware'. *"I distrust all systematizers and I avoid them. The will to system[atize], shows a lack of integrity."* Neurotic managers will find that ambiguity cannot be resolved into some tidy pattern, and that jumping onto a band-wagon of methodologies is merely impulsive stress-relief. The relief is a business-world full of insecure managers who feel that the remedies are successful for everyone else, but not for them.

This rash of solutions has itself become a problem! The ever-growing neurotic demand for answers, particularly technological answers, has closed the vicious circle, with even more 'solutions' causing ever deepening neuroses! In attempting to *Manage Change*, every organization will become neurotic! The best we can do is to be resilient in trying to cope with uncertainty!

Yet still, insecure managers believe that, through tidiness, first comes control and then success! But tidy classification is notoriously difficult, for it means attempting to set a boundary; a boundary that is chosen according to the human observer's particular purposes and priorities. The choice of boundary can be a source of doubtful classification; social systems are intrinsically ambiguous. For example, within an organization (a system), professionals (sub-systems) often form peer groups with individuals in other companies, perhaps within trade unions, to compare notes and experiences, and to optimize career opportunities. It is not uncommon that their loyalty is to their 'trade' (accountancy, computing) and not to their employer. And of course these individuals have family ties, political, philosophical and religious beliefs, so-called lateral loyalties, all adding to the ambiguity. Recognizing the inevitability of such ambiguity means that even the very identification of 'the organization' involves purpose-driven choice. Managers are left only with perpetually changing questions of appropriate or inappropriate choice.

This situation has caused me increasing concern, particularly as it dawned on me that I was part of an education system that was actively promoting misguided methods backed up by inappropriate technology. I was *"sanctifying so many lies."* As a scientist I was steeped in a culture of *"optimistic rationality"*, that by rational understanding things will get better. As a teacher I was expected to describe the world from a mixed perspective of mathematics, logic and morality. Yet the world I knew was beyond numbers, beyond true and false, beyond good and evil.

That last phrase should give you a clue as to how I was reconciled to this dilemma. The books of Friedrich Nietzsche, in particular *Beyond Good and Evil*, gave me a totally new outlook, a fresh confidence, justification and a mission. Although written over a century ago, the work of Nietzsche is of major relevance to today's managers, and so I will frequently refer to him during this talk.

Thinking Managers

From this perspective, and from the theory of systems, I propose an approach that could assist managers in understanding the opportunities and risks implicit in uncertainty. Rather than postulating rules, I shall merely suggest certain general principles that, suitably interpreted, should prove useful in practice. I would never recommend a specific approach, for generalizations are meaningless. The appropriateness of any approach is largely dependent upon the unique environment – its intrinsic singularity. There are no generic answers. Ultimately, 'thinking managers' must not shirk their responsibilities by delegating decisions to a rule-book. As the Ancient Chinese put it, *"when empires were doomed they had many laws"*.

Managers must *learn to live with uncertainty, and love it*, to enjoy confrontation with an uncertain reality. For uncertainty creates winners as well as losers; it separates the men from the boys.

The only sensible approach is to initiate plans, but to be flexible enough to react quickly to whatever risks or opportunities appear. A blinkered faith in planning, and using the past as a mirror to the future, is likely to constrain understanding, insight and lateral thinking of quality employees. Ours is an intrinsically singular, untidy world, where *"perfection of planning, is a symptom of decay"* (Parkinson). For *"to make plans, and project designs, brings with it many good sensations; and whoever had the strength to be nothing but a forger of plans his whole life long would be a very happy man: but he would occasionally have to take a rest from this activity by carrying out a plan – and then comes the vexation and sobering up"*.

Ultimately, it is the quality of 'thinking management', and that of their staff, to observe, interpret and innovate, to enjoy the responsibility for appropriate action and for conflict, that will be the major factor in taking decisions appropriate to the commercial environment. The people with the skills and confidence to handle this responsibility are in very short supply. Companies must value this 'human capital', perhaps more than 'financial capital'. No company can afford to lose such valuable human assets.

Those with the 'right stuff' require a stimulating intellectual climate, not a rule-based bureaucracy. Thinking managers prefer intellectual freedom to organizational constraint; they deny that procedure and process are valid replacements for action and decision. So emphasis must be placed on human resource management within organizations in order that staff are motivated to realise their full potential. Many compromises must be made to the present-day (so called) 'best practices' of business, if the stagnation caused by the underdevelopment or loss of this 'new aristocracy' is to be avoided.

The Disposition of an Organization

To prosper, an organization must evolve a *disposition* that will satisfy Ashby's *Law of Requisite Variety*: *"only variety can destroy variety"*. It must have the facility, be of sufficient internal variety, to confront and respond to perpetual changes in its environment, and hence to survive and flourish. A disposition to act is an implicit property of every organization, although it is in no way fixed and inflexible. Such a disposition is untidy – inherently ambiguous and constantly being refined and redefined via feedback from its interaction with the risks and opportunities around it.

The commercial environment is continuously changing, and to be effective an organization's disposition must be capable of dealing with it, winning from it. A company can only cope with uncertainty if it is *steered* by management with appropriate *tactical* responses. However, unknown future changes in the environment, will only be apparent in hindsight, and these have to be catered for *strategically*. Effective management steer their organization through its environmental turbulence, taking tactical advantage of the environment and avoiding risks, but only within the constraints of strategies that will enable it to deal with the unknown.

Strategy and Tactics

But what exactly do the terms 'strategy' and 'tactics' mean? Are they just more trendy buzz-words, like 'synergy', that are sloshing around in the international business community? The Oxford English Dictionary says that *"strategy is the art commander-in-chief; the art of projecting and directing the larger military movements and operations in a campaign. Usually distinguished from tactics, which is the art of handling forces in a battle or in the immediate presence of the enemy"*. Military strategy has been the source of numerous books from time immemorial, and the business world has zealously taken up the military metaphor. Most businessmen would agree with Napoleon's notion of strategy as *planned flexibility*, although they would shy away from his policy of selecting only the *lucky* generals. Tactics on the other hand are *"the arrangement of procedure, the action to be taken, in order to fulfil an end or objective."* They are not nearly as important; for tactics are what lieutenants do, but strategy is the work of generals!, a most peculiar attitude since it is the appropriateness and quality of tactics that separates success from failure in the short term.

Let me now introduce my own use of the words *strategic* and *strategy*. I use the adjective *strategic* to describe any influence that has a lasting effect on the disposition of an organization. This influence will have emerged from human actions, but not necessarily human designs. It will be perceived as a benefit if it increases the organization's ability to generate variety (that is to be resilient), and a loss otherwise. I consider a *strategy* to be purposeful human decision and action that is intended to be strategic.

But, not all strategies are strategic; and not all strategies achieve their intended strategic aims. Not all strategic effects are the result of strategies – they can be the consequence of unintended actions. There can be no guarantees – with its disposition, an organization may prosper in its environment to a greater or lesser extent. Failure may mean eventual extinction.

Time constraints cannot be imposed on a strategy – the time factor only refers to the duration of the effect on the disposition. Thus, strategies should not be equated with long term plans. Neither should they be identified with beneficial changes, but only with intended change, and that may not be achieved.

Strategy is not the sole domain of management; it is not just purpose expressed by the powerful. The actions of everyone in the organization will influence its disposition to a certain extent. Strategies may increase or decrease internal variety, ultimately they may or may not achieve success for the organization; a strategy is not a panacea. The emergent disposition of the organization or its environment may finesse and even diametrically oppose the original intentions of the strategists.

Design only works if it is not overly contentious to the present disposition. We cannot just impose preconceived notions of 'generic' strategies onto the apparently blank sheet of an organization. Organizations have complex predispositions even before the strategists start, and these dispositions will evolve and emerge from human actions, with or without, even contrary to, strategists. Misapplied and inappropriate strategies may result in tension, disruption, and both constant and perpetual change within an organization, with highly detrimental effects. But neither must they be bland – *"no pain, no gain"*.

By its disposition, an organization attempts to profit from its environment. It anticipates; it senses, interprets, and reacts to events observed in the environment. Data on these events are transmitted to the control centres of the organization. There the management treat it as information, not mere data. Management can then formulate forecasts of trends, model the data, derive *plans* and transmit *tactics* back to organization's components, in order to take advantage of the circumstances. It is innovation that affords a greater store of tactics, and variety that is the potential to innovate. Competition is tactical, the ability to compete is strategic.

My interpretation in no way diminishes the importance of tactics, since the immediate survival and expansion of an organization will depend directly and essentially on tactical success. The tactics used to achieve a competitive advantage may coincidentally turn out to be of lasting effect on the organization's disposition. However, whether this strategic influence will be beneficial is by no means certain. It is perfectly conceivable that the tactically advantageous approach, of concentrating on previous successful procedures, will reduce internal variety and so restrict future capacity to innovate.

All tactics become unquestioned ritual, and lead to crisis, if they are not abandoned in time. Exactly because they are meant to exploit confidence in the circumstances of an enterprise and prescribe specific and potentially successful moves, they falter as soon as doubt springs from a change of circumstance. If tactics are continued despite change, they effectively prevent the manager from being innovative, and introduce an organizational inertia that ignores new environmental conditions.

In addition to their inertial influence on decision-taking, tactics may allow new problems to emerge. These can, even over a short period of time, acquire such prominence, that managers are compelled to concentrate much of their energy on the tactics themselves, which were designed to support them in their enterprise.

The Transitory Nature of Tactics

My interpretation of strategy, not as a plan or method, but as an intention to cause a lasting effect on an organization's disposition, should be taken as a matter of common sense. Naturally tactics, plans and methods have their place in day-to-day management and administration. However, actions under conditions of uncertainty are essentially where the tried and tested methods do not inspire enough confidence in the application of tactics.

Given the uncertainty and continual change in the environment, any tactical gain will be transitory; consequently, changing trends will require changing tactics. The more variety in the business environment, the less control the organization has over the course of events. The ability necessary to reformulate subsequent perception and tactics, in order to combat these environmental changes, will require much internal variety.

A resilient environment has a potentially unlimited variety of moves in store. This demonstrates the paramount importance of resilient strategic management over short term tactical manoeuvring. Without continuous innovation and generation of the required internal variety, in the long term, an organization will be in a permanent state of crisis management.

But just as tactics cannot fulfil a strategic task, strategy does not replace tactics in the aggressive drive for short term gains. In order to formulate tactics, it is necessary for management to choose, observe, systematize, measure and interpret the environment. This will involve the mapping of observations onto analogies and models based on previous personal experience or onto formal methods – to be tidy to a certain extent. But tidiness is not truth. Managers must never forget that each system, measure, model, analysis and analogy is based on choice. Each is a distorting and distorted filter. Every model will involve arbitrary and simplistic measures, comparisons, classifications, and syntheses.

Models ignore the untidiness, the debris of detail in the unfolding history of the organization being modelled, and in its environment. These memory fragments, when reconstituted as opportunities and risks by a particular contextual significance and relevance, have the potential of changing the disposition of organization and environment.

In their models, designers can only see a tidiness that is a limited snapshot of an ordered functionality and utility. Although we have no alternative but to use this approach, we must nevertheless be aware of, and question, its appropriateness, relevance, flexibility, validity, consistency and permanence.

'Dispositioning'

Unless the strategic context of information is fully appreciated, and unless managers are aware of the true nature of crisis, complexity and uncertainty that awaits them, then strategies will become mere wishful thinking, and tactics will fail. Management is about taking risks. Grasping the place of management within the business reality is a never-ending process. Ashby's Law of Requisite Variety implies that a manager should ensure that his organization can generate enough internal variety of responses to be able to react tactically in an appropriate manner to changes in the business environment. This calls for a form of strategic management that I call 'dispositioning'. Dispositioning involves a recognition of the need for short term tactical thinking, and the preparation of resources within an organization, enabling it to function and to contend in a preferred way with, not only known commercial competition, but also the uncertainty, complexity and ambiguity inherent in an unknown and unknowable future.

Success is more likely through a strategy of keeping systems small – "*small is beautiful*" – small is controllable, small is flexible. Flexibility means new tactics can be developed as and when needed to take advantage of the environment. Flexibility means a willingness to reverse any commitment to historically successful procedures. Flexibility means not trying to impose a bureaucracy of tidiness on an untidy commercial world. Of all organizational resources, the work force is the most flexible, and that flexibility should be encouraged and exploited through the provision of education and training schemes. 'Management by learning and experiment' is crucial. Learning and experiment involves a sense of irony, a recognition of the perverse arrival of circumstance, that mocks any insistence on a measured tidiness. Too often there is a pseudo-scientific refusal to accept that consequences of actions are not correlated with intentions. The perversity of consequence turns design logic on its head. For paradoxically, even though 'solutions' may have a questionable theoretical basis, they can nonetheless prove extremely useful. "*The falseness of a judgement is to us not necessarily an objection to a judgement...*" Only after accepting this unscientific yet opportunistic insight will the search for a tactical payoff give major commercial returns. The steering of an organization must be grounded in perpetual observation and experimentation, in contingencies, and in a sympathetic reaction to the disposition of the social and commercial environment. Learning from mistakes; but there must be no obsessive compulsive neuroses about failure. This must be a sceptical approach, not based on a naive ritualistic belief in a description through measurement. Continuous experimental feedback is essential within each organization, in order to cope with unpredictable, unintended consequences of system behaviour. Commercial success will only be found if the industrial and educational infrastructure is expanded beyond the utopia of naive technological systems and inert theories, to a sympathetic understanding of what is appropriate and inappropriate.

It takes experience, not expertise, to come to terms with dynamic systems and with the human and commercial environment. There has been far too much emphasis on rationality over the past century of management theory. Economists insist on 'rational expectation', that we explain our mistakes so that systematic and persistent errors do not occur. Bunkum! The world is not rational. "*Rational thought is interpretation according to a scheme that we cannot throw off.*" "*The world seems logical to us because we made it logical.*" Our world is perpetually changing, there are no explanations only transitory descriptions. Our descriptions, words, are where confusion begins.

Management wisdom tells us that we must be 'nice', and that 'stress is bad'. Yet more bunkum! The variety needed to succeed will generate conflict. Of course this can be mitigated to a certain extent by a predisposition within an organization towards alertness, patience, tolerance, perseverance, imagination, innovation, inventiveness, commitment, motivation, productivity, experience, quality and many of the other unmeasurable human virtues. But virtues based on strength, not weakness. "*I have often laughed at the weaklings who thought themselves good because they had no claws.*" There has to be a disposition to confrontation – that even welcomes it. Only through conflict will systems be tempered in the flames of commercial competition. "*Paradise lies in the shadow of swords.*"

Decision-taking under uncertainty is a matter of a manager's perpetual analysis, of the organization, environment and himself, it is a personal affair. A strategy that supports an understanding of uncertainty has to advocate flexibility, variability and strength. This means that a manager must shoulder the responsibility of excelling individually. All actions are situated, and only in a manager's self-determined reality do they acquire their meaning. In any other context they will acquire a different

interpretation. Thus, a manager has to steer in the flow of events as they appear meaningful to him, and thereby take risks. Dispositioning recognizes that there is no virtue in a common vision. For a common vision is not necessarily a sensible vision, it is merely ... common!
Leadership is called for; not leadership that panders to the 'greatest good' or egalitarian ideals, for that is a formula for mediocrity. *"There is no justice in equality."* To succeed, leadership must be strong, ruthless, in its goal of variety. *"Nature is not immoral when it has no pity for the degenerate"*; degenerate, that is, in the sense of reducing an organization's capacity to generate variety. Healthy systems expel all such poisons from within.

Facing uncertainty is a matter of welcoming that the unimaginable can and will happen, of being prepared to deal with it on the level of personal choice, of accepting responsibility, with a sense of wonder and positive thinking, and with strength, maximizing opportunities and minimizing risk.

An Introduction to Project Management

Angeliki Poulymenakou, Department of Information Systems, London School of Economics and Political Science, London, United Kingdom

Introduction

Project management can be presented and discussed from a number of perspectives, however, no single perspective yields a 'complete' picture. People, organizations, techniques and technology are all part of project management, and yet none of them is sufficient on its own for achieving the purpose of the process: the controlled execution of a predefined piece of work.

Project management is a 'live' topic, and its profile is changing to reflect the current status of management practices. Understanding project management helps us to make sense of the complexity, and to reduce the uncertainty, in sustained organizational activities. Organizations need to respond rapidly to changing conditions in their environment, and individuals in them need to adapt their behaviour and work practices in view of this constant process of change.

Projects are managed by people who apply knowledge, skills and experiences, who pursue organizational and personal objectives and who make and maintain assumptions about the world in which they operate. In this article, aspects of the project management situation are presented that touch upon social, organizational, technical and technological issues. In some cases specific courses of action are described, whereas in others, issues are presented and discussed. The aim of this document is to introduce, rather than to exhaust, the project management topic. Any reader wishing to 'experience' project management, should realize the need to read more, to think more, and of course, to do more.

What is a Project?

*A unique, well defined effort to produce **specific results**, in a multi-functional environment, with a **set time frame** and at a **given cost**, employing a multi-skilled team under special management.*

Project Management

With its origins in the practical management of large engineering projects, project management is basically a systematic, structured approach to uncertainty. Projects are developed in an environment often characterized by the risks of change and failure. These risks vary greatly between high risk projects, where the chances of stability and success are limited but the prize at the end is considerable, to low risk projects, where the end result is relatively predictable. Development projects are typically more likely to succeed than research projects, where the final deliverable is less certain.

The broad concept of a project, and its associated project management, is applicable across a wide range of industries and situations, from physical production to administrative activities; examples include construction projects, the production of an annual report, new product development, the relocation of a company, and even building a house extension. In terms of human resource management, good project management can engender a considerable sense of commitment by the project team to the successful completion of the project. Perhaps even more than other forms of management, project management is above all a practical skill, gained mostly through experience.

Types of Projects

One way of distinguishing among projects is in terms of the industry in which they are undertaken; in this sense there are projects in construction, engineering, public administration, military, manufacturing, information technology, etc. Another possibility is to look at their size, time scale and budget; in which case projects can be characterized as small, medium and large. The metrics accompanying this distinction vary substantially across industries: for example, in the telecommunications industry a small project has up to 5 people, takes up to 6 months, costs up to $100K; a medium-size project has up to 15 people, takes up to 3 years, costs up to $3M; and a large project is anything beyond those limits. Finally, projects can be characterized in terms of the nature their function and orientation of their objectives. In this sense, there are *application development projects* that use existing infrastructure, resources, products and processes in an organization (e.g. the construction of a bridge uses well established resources and follows known procedures); *infrastructure development projects* where either new or updated practices or standards are introduced as a result of the project (e.g. setting up a new research institute); *product development projects* (e.g. building automatic cash dispensers to complement manual cash handling in banks); *research projects* that are exploratory and open ended in nature; and projects that display a combination of the above characteristics.

A project should have a definite start and finish and it thus unlike the continuous or repeated processes suited to traditional line management.

Projects, People and Organizational Environments

Any project represents a scaled down version of the social, organizational, business, technological and political environment in which it takes place. Every project is undertaken in an **intersection of these environments**, and their characteristics influence the activities, progress and products associated with the project. The most important attribute of a project are is the **people** in it. They carry out the activities, set the goals, and justify the existence of the project. The fact that any project is taking place means (or should mean) that somebody needs results. The people with an interest in a project, are its *stakeholders*. These include the members of the project team, the managers, the clients, sponsors, contractors and users as well as anybody who is directly or indirectly influenced by the existence or outcome of the project. Different stakeholders develop different interests and concerns in a project. The clients and users focus on the end-result, the team is concerned with the work content, whereas the sponsors are preoccupied with the time and budget attributes of the project. It is the project manager's responsibility to accommodate, reconcile and satisfy all these different views of the project; this requires the application of **special management techniques** as well as a **variety of skills** from the project manager.

Role of the Project Manager

He (or she) has to plan, organize and manage the project in order to produce the desired results on time, within budget, in accordance with its objectives and to the satisfaction of all the stakeholders. The project manager is the link between the project and its environment. Given the different interests people have in a project and the variety of interrelationships among stakeholders, conflicts in requirements and approach are rarely avoided in a project. Moreover, since any project operates in the intersection of so many environments, changes in any of them may change the project's own requirements or objectives. Therefore, the project manager's job is invested with **complexity**: he has to work amidst conflicting interests in, and perceptions of, the project, and to cope with changes to those over time. The fact that the project manager has to adopt the viewpoints of the various stakeholders, speculate on their expectations, gauge their intentions and anticipate their actions, introduces great **uncertainty** in his job. This uncertainty is further accentuated by the project manager's own assumptions on the productivity, amenability and capability of the members of the project team.

A recent study of project managers suggests an order in which they prioritize the challenges to their management skills in projects (Thamhain & Wilemon, 1986):

coping with end-date driven schedules;
coping with resource constraints;
communicating effectively among task groups;
gaining commitment from project participants;
establishing measurable milestones (key project dates);
coping with changes;
securing an agreement on the project plan within the project team;
securing senior management commitment;
dealing with conflict;
managing external participants (vendors and subcontractors).

Why Project Management?

A project, as a piece of well defined and controlled collective effort, has a number of dimensions along which it needs to be managed:

The technological aspect has to do with the feasibility of achieving the project's objectives in terms of currently available procedures and tools.

The organizational aspect is related to the division and management of the workload, the provision and distribution of information amongst project participants, the patterns of communication and the utilization of resources.

The historical aspect places the particular project in the wider organizational setting of other projects, organizational goals and means for achieving them.

The human resources aspect of a project refers to the group dynamics of the project team such as the willingness to co-operate and distribute roles and responsibilities inside the team. This aspect also refers to the individuals in the team in terms of their talents and skills, experiences and motivation.

Last but not least, the political aspect of a project refers to the vested interests and power struggles in the organisation to which the existence and 'fate' of the project give rise.

Problems in projects manifest themselves in a variety of ways. In most cases, however, the first sign of trouble is a time delay followed closely by overspending. Experience in many fields (engineering, third world development, military, software) suggests that, left to their own devices, projects are likely to drift indefinitely. Recent figures suggest that 50 per cent of software development projects go 100 per cent over their allotted time and budget. *Parkinson's Law* states that *work expands so as to fill the time available for its completion.* A special type of management is needed to utilize expensive resources (people, equipment etc.) both efficiently and creatively. There is a need, within a complex uncertain environment, to answer the questions:

> How long will the project take to complete?
>   How can the schedule be shortened?
> What resources are required for the project?
>   How many? For how long?
> What are the resource costs?
> What is the effect of a delay in a particular activity?
> Which activities are critical to the project schedule?

The Project's Manager's Work: Goals, Activities and Information

Three fundamental goals underlie all actions, decisions and deliberations that the project manager engages in throughout the duration of a project (Berkley *et al*, 1990). The project manager needs to *transact well with the project environment, manage well the work on the project* and *advance his personal and his team's experience and reputation* through the participation in the particular project. The first two of these goals need further decomposition. In interacting with the environment, the project manager needs to *clarify the requirements set on the project* by obtaining the views of the various stakeholders and reconciling the differences that may occur in these views. Everyone involved in the project should develop and maintain the same perception of the objectives of the project as well as means that will be employed to meet these objectives. The project manager is also responsible for *satisfying the clients' and users' needs* in the project. This is achieved through an on-going negotiation with these agents both on the basic set of needs and requirements and on subsequent changes to them. In order to manage the work on the project, the project manager needs to *create and put to use a project plan.* This plan covers all the activities, tasks, people, resources, events, results dependencies and risks associated with the project. The project manager is also responsible for *maintaining a relationship between the project plan and the reality* in the project throughout its duration. Finally, the project manager is responsible for *facilitating project work* through the provision of feedback to and from the other stakeholders.

Activities

A great leap forward is best accomplished in short, comfortable hops.

In order to handle the complexity and uncertainty inherent in any project, the project manager needs to *decompose the project* into manageable parts and *assess the risks* associated with each of them. In terms of *time*, a project is broken down into **phases (or stages)** (and perhaps sub-phases), and then *work* in each of them is broken down into **tasks**. Large, complex projects may require further levels of decomposition. The major phases in any project are:

taking over;
initial planning;
launching;
running;
closing.

The forms of this generic phase structure tasks vary across projects. For a construction type of project, the major phases are initial planning, feasibility study, construction, and testing. For a specific project, the phases should provide a clear description of the major parts of the work that needs to be done. Consider, for example, this project: building an extension for your house. The phase and task decomposition for this project is shown in Figure 1.

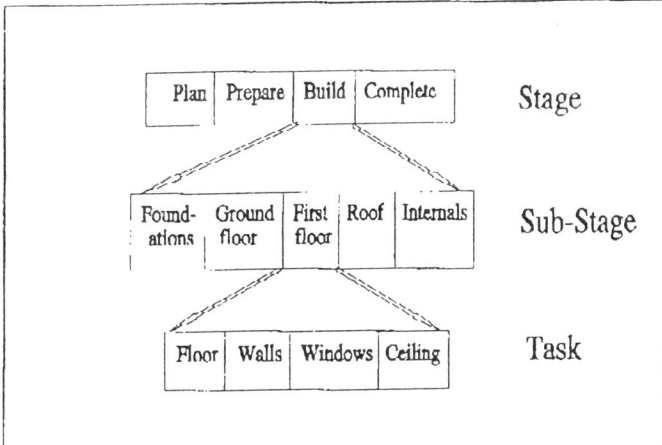

Figure 1. Project stages, sub-stages and tasks for a house extension.

Project Stages & Deliverables

The stages are the major phases of a project used to identify different groups of activities, to mark changes in the nature of and advances in the work undertaken and to declare points in time for the duration of the project (milestones) where work delivers a specifiable and tangible output (deliverables). At the end of each stage, the end of stage review acts as an important control mechanism. The deliverables may be related to the product (e.g. particular components) or the development process (e.g. plans, assessments, decisions). These deliverables should be suitable for undergoing quality assurance checks. They should be complete and understandable, represent the natural consequence of the previous stage, and the natural forerunner of the next stage. Activities in a project belong to three major categories, which necessarily overlap:

a) Planning
   This is carried out at two levels: the first, an overall plan for the whole project, which should be complete before the project starts, and which can be used for project justification etc. The second, more detailed level, is produced as the project progresses and involves the detailed plans for the current and next stages. The planning procedure comprises the definition of tasks, their duration, sequence and dependencies. The resources required have to be estimated and then allocated.

b) Monitoring and control
   This involves a comparison of the current position against the plan. Divergences should be identified and corrected and the plan modified. This process is a continuous one, with regular reports and reviews.

c) Risk management
   This involves the assessment and monitoring of risk in a project. A risk, is a reference to the occurrence of an undesired event. In other words, a risk is a reference waiting to happen. There are managerial and technical aspects to risk. The management of risk has *proactive* (i.e. acting before a risk transforms into a problem) and *reactive* (i.e. rectifying a problem once it appears) elements to it.

Project Planning

It is hard to underestimate the importance of project planning – *If you can't plan it, you can't do it* and *Get it right, up front*. Poor planning in the early days of a project is a frequent cause of project failure, but it may not show up until the later stages. At the same time, we must recognize that planning can only attempt to control uncertainty, and, rather than trust to a deterministic model of the future, we should allow some contingency and flexibility within the plan. As such, planning is a key problem of project management and it is doubly important as the plan forms the basis of project monitoring and control. The normal approach to planning is as follows:

Identify stages:
    Main project structure
Identify tasks, sequence and dependencies:
    What has to be done
    When it has to be done
Estimation:
    How many/much resources for **how long**
Resource allocation & scheduling:
    Which resources
    How they are to be deployed
Risk assessment:
    Identify risk drivers.

Plan Information

The project plan should clearly show:

Stages and tasks:
    network charts and resource schedules (for people and equipment)
Milestones (checkpoints) at the end of each stage
Constraints:
    objectives, cost, time, staff resources,
    co-ordination with other projects
Deliverables for each milestone.

Estimation

As part of the planning process we need to estimate the amount of work to be done and the time (person-days) to do it. This is the *Achilles heel* of planning as there are few satisfactory methods available. It is often a judgement of: *how long is a piece of string*? The most reliable method is the *black art* of professional judgement (rule of thumb). In most cases, estimation depends fundamentally upon experience (and luck). In some cases, where there is a history of similar projects, estimation can be based on a comparison between the current project and earlier ones. In certain, highly structured situations which can be modelled mathematically, quantitative techniques may be relevant. However, such techniques do not allow for the unpredictable variances encountered in organizations. In practice, the selection of methods depends on the type of project and the degree of uncertainty.

Despite the difficulties, estimation should be carried out carefully, in order to prevent the common phenomenon of *GIGO – Garbage-in Garbage-out*. Estimation should be performed at an appropriate stage (not too early/late in the project) and at an appropriate level of detail (not too detailed/abstract). Project managers are advised to consult with the whole project team, but the final estimate should be that of the project manager – never estimate in committee, nobody will admit responsibility. Estimates should be based on the performance of average team members, and not on the best staff. Ideally (although there is rarely time available) managers should monitor their success in estimating; the learning is worth the pain.

Resource Scheduling

This is the process of allocating resources to tasks. The duration of many tasks depends upon the amount of resources supplied; for example, 10 man-days is sometimes (not always) equivalent to 10 men for 1 day or 1 man for 10 days. Clearly, resource scheduling has to be related to the availability and cost of resources. There is usually a time-cost trade-off, such that the project can be completed earlier, but at an increased cost. Cost is not usually a continuous function due to higher overtime rates and varying overhead (indirect) costs.

Monitoring

This is the systematic comparison of the progress of current and completed tasks against the original baseline plan (and the most recent reschedule). It should form a continuous record of the project's progress and the results reported to the sponsors/users, as well as being fed back to the project team. The purpose of this progress check is to determine quickly when actual progress is deviating from the plan. In general, monitoring should always look towards the future, such that the project manager examines the current performance in the light of its impact on the subsequent tasks and stages. However, at the same time, a complete project history is useful for purposes of comparison and organizational learning. Monitoring is primarily based on observation, reporting and the analysis of the results. In a large, complex project, the project manager should focus in particular on the progress of the critical tasks as these are the ones most likely to delay the project.

Control

Many tasks (especially the more interesting ones) tend to be self-perpetuating rather than self-regulating, and therefore they need the external control provided by the project manager. *Left alone, projects will continue indefinitely, roaming like lost sheep through the wilderness for years.* The following are essential elements for effective control:

Clarity of objectives and priorities;
Accurate and timely reporting;
Identification and communication of problems;
Flexibility in terms of keeping options open as well as maintaining a contingency of spare resources.

An important part of the control function is the *end of stage review*. They should involve the project manager, the project team, sponsor/users and other interested parties (e.g. auditors, senior management). These reviews should consider the following information:

Completeness and quality of the product;
Status of the project;
Overall plan for the rest of the project;
Detailed plan for the next stage.

The deliverable of the end of stage review is the decision to either:
Continue with the next stage;
Repeat the current stage because progress/product was not satisfactory;
Pause for a while with the project, while further investigations are carried out;
Abandon the project.

Although it is a difficult decision, it is often wise to abandon a project that is not producing the desired results. It takes a brave project manager to recommend abandoning *his* project, but the saving in further investment from the organization should not go unrewarded.

End of task reviews are much smaller versions of the above, involving just the project manager and the team members concerned. Depending upon the structure of the project, regular reviews or reports on a monthly basis are useful where the project stages are large. Such regular reports should complement the stage and task structure.

Steering Committees (Project Boards)

For medium and large projects, a formal steering committee may be required to oversee the project. The committee should be made up of representatives from all interested stakeholders, including users, sponsors, the project team and any other technical or non-technical departments that are involved. The level of representation (senior management, middle management etc.) should be determined by the nature of the project and the functions of the committee. For very large projects, a fairly senior team is required, to whom the project manager can report. However, the functions of steering committees can vary considerably between projects and organizations, including:

Strategy formulation;
Resource allocation;
Performance review;
Forum for discussion;
Approval of major projects.

Whilst each function is valid in its own right, the functions may not mix, and great care has to be taken regarding the terms of reference of a committee. Although such committees promise considerable benefits from the participation of senior management and the involvement of the various groups, there are certain issues that have to be resolved satisfactorily. These include:

Is the committee a decision-making body, a 'talking shop' or a working group?
How should members be selected?
What level of seniority or expertise is required?
How often should the committee meet?
Who sets the agenda? Who sees the minutes?

Components of Projects and Techniques for Project Management: Tasks and Resources

Tasks are small, manageable 'chunks' of work, of which there may be two or three hundred per project. Each task should produce an identifiable deliverable, which is specifiable in advance and can serve as an end-of task checkpoint. Each task should also be small enough to control easily (perhaps no more than 10 man-days work).

Tasks come in a wide variety of types including:

Planning and scheduling;
Execution and reporting;
Response, change and control;
Training and documenting.

In order to manage (plan, monitor and control) these tasks, we need to determine the following information for each task:

Unique task identifier;
Duration length (days, hours etc.) and whether it is fixed or variable;
Start & finish dates:
    planned (according to the original 'baseline' plan)
    scheduled (according to later rescheduling)
    actual
    constraints
Resources required;
Relationship with other tasks:
    predecessors & successors.

In addition to the tasks, we also have to consider the **resources** employed on the tasks. Resources are assets of value (or cost) to the organization, and which usually have some opportunity costs in use. There are typically a large variety of resources, including:

People – staff, managers, contractors;
Equipment – purchased, constructed, leased;
Other – buildings, capital.

These resources can usually be employed in a number of different ways and so we are concerned with their efficiency (*doing things right*) and their effectiveness (*doing the right things*). Resources require careful management in order to maximize the benefit to the organization.

For management purposes, the following information is required for each resource:

Unique identifier;
Maximum quantity available;
Working hours;
Costs;
Tasks to which resource can be assigned;
Other projects to which resource is assigned (resource sharing).

Work Breakdown Structure (WBS)

This is a hierarchical project component identification schema, which reflects the stage structure of projects. It is used to decompose a project according to the major categories of work included in it. For example, our house extension might be depicted as follows:

Plan for the extension:
    Design the extension
    Hire builders

....
Prepare for the extension:
    Prepare the grounds
    Demolish existing structures

....
Build extension:
    Construct foundations

    ....
    Construct first floor:
        Construct first floor flooring
        Construct first floor walls

    ....
    Construct roof

....
Complete extension.

Critical Path Method (CPM)
Program Evaluation and Review Technique (PERT)

There are two closely related techniques that are often combined together to form the basis of modern project management. However, they originated independently: CPM was developed by the Dupont Corporation in the 1950's to assist project scheduling for plant overhaul, while PERT was developed about the same time by the US Navy for scheduling their Polaris missile project. We shall use the term PERT (as this is the term found in the software). The function of PERT is the detailed, bottom-up planning and scheduling of projects that comprise a large number of interrelated tasks. The method involves producing a graphical representation of a project in terms of a network of tasks and their dependencies (for example, task A must be completed before task B can be started). Examples for the house extension project are given as Figures 2, 3 and 4. We shall follow the common practice of representing tasks as nodes of the network and dependencies as arcs.

Procedure

1) Identify all tasks and their dependencies;

2) For each task:
    Estimate duration
    Calculate earliest start/finish times:
        forward pass through the network (partially ordered sort)
    Calculate latest start/finish times:
        backward pass through the network
    Calculate slack time = Latest start − earliest start

3) Determine critical path:
    The path made up of the longest sequence of dependent tasks
    Tasks for each slack = 0.

In general, if the critical path is shortened, the project will finish earlier. The critical path may be modified in the light of the availability of resources. For simple projects, the analysis can easily be performed manually, but computer support is required for more complex projects.

A Simple Example

Two people changing a wheel (adapted from Taffler, 1979). Table 1 shows the tasks, the resources required (in terms of which person will perform the task), the task duration (in minutes), predecessor tasks, earliest start time (EST), latest start time (LST) and the available slack time. The same schedule is shown as a PERT chart in Figure 5. As can be seen, the 'assistant' (person B) is the only one with slack time.

Task No.	Description	Person	Duration	Predecessors	EST	LST	Slack
1	Jack up car	A	10	-	0	0	0
2	Remove wheel	A	5	1	10	10	0
3	Take out spare wheel	B	5	-	0	10	10
4	Put away old wheel	B	3	2,3	15	22	7
5	Fit spare wheel	A	5	2,3	15	15	0
6	Release jack	A	5	5	20	20	0
7	Finish		0	4,6	25	25	

Table 1   Changing a wheel

Approaches to Shortening the Critical Path

Shortening the critical path is the key to earlier project completion
The critical path is the set of critical tasks for which any delay will delay the whole project and so the critical path, in a sense, 'determines' the project completion date. Thus, project managers should pay particular attention to try to shorten the path. Possible approaches are:

> Add resources – at a price!
> Review tasks on the critical path – try to shorten them or move them off the path:
>> Relax constraints and dependencies – increasing the known risks
>> Break down critical tasks into smaller tasks that can overlap, increasing the number of dependencies and the overall complexity.

General Hints

> Keep high risk activities off the critical path – the risk often implies delay and complexity;
> Keep the project manager off critical activities – the project manager needs time for managerial tasks;
> Keep inexperienced staff off critical activities;
> Try not to schedule team members to carry out more than one task at a time.

Result of PERT Analysis

The PERT analysis produces an ordered set of tasks and dependencies and a first approximation of:
> Earliest completion time for the project;
> Earliest/latest start time for each task;
> Earliest/latest finish time for each task.

It allows the project manager to identify the critical path and the location of slack time, some of which can be kept in reserve as recovery time. It should be noted that there are two types of slack time:

a)   Total slack
    The time by which a task may be delayed without affecting the project finish date.

b) Free slack
The time by which a task may be delayed without affecting other tasks.

Advantages of PERT

This approach is widely used because:

It provides a systematic basis for planning and scheduling;
It is a relatively simple way to identify task dependencies and the critical path;
Graphical representation facilitates communication;
Flexibility facilitates monitoring and rescheduling;
Useful for 'what-if' analysis.

Disadvantages

Although it normally forms the foundation of project management, PERT analysis has the following drawbacks:

Network diagrams are often highly complex;
It depends crucially on the accuracy of estimation;
It cannot always represent the subtleties of the real world.

Advances on Basic PERT

There are many variants on the basic techniques. For example the network can be represented as a time-scaled network, where the tasks are depicted against a time scale in terms of days, weeks, etc. The 'true' PERT technique (as opposed to CPM) attempts to cater for non-deterministic tasks and durations by using scheduling based on multiple estimates:

optimistic (minimum) time;
pessimistic (maximum) time;
most likely (modal) time.

Gantt Charts

The other common analytical tool is the Gantt chart, which traces its origins back to the First World War. This is a bar chart that shows tasks against a time scale, such that the length of the bar represents the duration of the task.

An example of a Gantt chart for the house extension project was given as Figure 2. These charts are particularly useful for highlighting task durations and the parallelism of tasks. The whole project can be depicted against a calendar, with the critical path clearly marked. It can also be used as a tool for monitoring, showing the percentage of each task has been completed.

Issues and Trends in Project Management
Project versus Line (Functional) Management

Clearly, the importance of the project management approach varies considerably between organizations and industries. In some cases (e.g. software development), most of the work is project based, while this may be the exception in other industries (e.g. retailing). From a management perspective, project management is far from easy. However, the notion of a project team of various specialists, led by a distinct project manager, has certain inherent advantages compared to managing projects using traditional line/functional management. In the latter case, the specialists are grouped together into functional units (e.g. marketing, systems analysis) each with its own management structure. The advantages of project management are as follows:

Improved project control, especially in complex, high risk environments;
Faster decisions, in the face of considerable uncertainty;
Improved accountability through the role of the project manager;
Improved team motivation;
A tendency for team members to gain an all-round knowledge (becoming generalists).

On the other hand, the project management approach has a number of inherent drawbacks, compared to line/functional management:

Senior management may lose track of projects as the responsibility is concentrated of the project manager;
It is hard to realize economies of scale;
The redeployment of specialists between project teams may be difficult both during projects and at their completion;
It becomes harder to develop specialist skills within the organization;
It is harder to apply standards;
The career structure for team members may be less satisfactory.

For industries where project management is the norm, a form of matrix management is often employed. Under this schema there is an overlap between project management and functional management. The functional managers maintain a pool of specialists and allocate them to projects to work under project managers. The drawback of this arrangement is that the specialists have two 'bosses', with conflicting objectives, resulting in the dangers of divided authority.

Reasons for Project Failures

As noted earlier, projects often fail. There are myriad reasons for failure, depending upon the unique circumstances of individual projects. However, there are some common threads clearly visible across different industries. The list below, taken from Gildersleeve (1985), refers to IT projects but are also common elsewhere:

Failure to identify the user;
Research confused with production;
Firm commitments made on the basis of inadequate specification;
Failure to obtain user approval to continue;
Failure to agree with users about the basis of acceptance;
Necessary tasks overlooked;
Task dependencies overlooked;
Project manager overburdened with detailed tasks;
Plans making no allowance for contingencies;
Checkpoints were not used to monitor progress;
Performance was not adequately controlled;
Communications broke down etc.

Failing Projects and Corrective Action

The symptoms of a failing project are fairly clear: the project starts to run late, costs escalate, and there is a general feeling that the project is slipping 'out of control'. The corrective action that the project manager *should* take to minimize the delay and cost over-run is equally clear. He should rework the plan, taking advantage of any slack time. He should review the critical tasks, perhaps breaking them down into simpler overlapping tasks, as well as reviewing the constraints and dependencies in case a reordering of the tasks can result in a faster completion. A useful tactic is to reallocate the better (higher performing) staff to the critical tasks, because in many occupations the difference between the individual performance of the best and worst staff may reach 10:1. The redeployment of other resources may also prove to be helpful. When projects start to run late, it is tempting to panic and draft in additional help without thinking. In some cases, this can solve a short-term problem, but experience in software development projects is largely the reverse. This phenomenon gave rise to *Brooks' Law* which states that *adding manpower to a late project makes it later*. This is because the effort required in additional training and co-ordination can exceed the contribution of the extra staff. An important factor that is often neglected is the effect of a failing project on the morale of the project team. It is essential that motivation be not just maintained but increased – this needs careful handling.

Whilst, in the classroom, it is fairly clear the action that the project manager should take, in practice, project managers (especially inexperienced ones) tend to forget the benefits of the systematic approach in favour of 'cutting corners' for short-term advantage. Thus, they will typically skimp on planning, documentation, work standards, and quality assurance. They may also succumb to the temptation of neglecting the management function in favour of carrying out the critical tasks themselves. Such actions have predictable consequences. With reduced control, the project may lurch completely out of control with the costs and delays increasing exponentially. The project then becomes a prime candidate for abandonment. If the project is completed, the final product may well be incomplete or faulty. This then requires immediate changes or repairs, but such modifications are severely hampered by the lack of documentation and the poor workmanship.

Trends in Project Management

A clear trend is the increasing formalization of project management. This can be seen in the growth of project management methodologies; for example, PRINCE (Projects in Controlled Environments) has recently become the UK government's standard methodology for IT projects. Similarly, there is a continued interest in the measurement and qualification of projects through project metrics. This is facilitated by the increasing use of automated support tools, such as Project for Windows. Whilst this increasing formalization and systematization has certain benefits, a brief examination of the key problems of project management will show that this approach only addresses *some* of the problems.

Fundamental Problems of Project Management
The Mystique of Quantification

The sophistication of network drawing techniques (e.g. CPM/PERT) and supporting software (e.g. Project for Windows) can lead to a mistaken belief in the accuracy and determinism of the resulting plans. Considerable faith can be engendered in the accuracy of the quantities of work and resources manipulated by these techniques, leading to the illusion of control.

However, many projects take place in a commercial, technological and social environment typified by uncertainty and change, where estimates and predictions are almost certain to be wrong and plans are little more than general guidelines. In these cases, flexibility and awareness will always triumph over mindless quantification. Another example of this unfortunate tendency is the quantification of risk. Risk analysis, the identification of potential dangers and the setting up of contingency plans, is a valuable strategy. However, where spurious statistical probabilities are allocated to individual risks, we have the dangers of the mystique of quantification.

Project Development is a Learning Process

Often the project team has limited knowledge of the problem and the problem domain. The rest of the organization may be similarly ignorant regarding the project development techniques used by the team. Thus, the project team (and any participating departments) have to undergo a learning process. If the project team is fortunate, it will include a 'project guru' who can combine a large amount of domain knowledge with good communication skills, so as to help motivate the other team members. However, all too often this process takes a considerable amount of time and may be characterized by errors. *"Producing the final deliverable isn't the problem, understanding is the problem"*. Project managers need to be aware of the learning process, and need to make sufficient allowance for it in their plans.

Project Development is a Political Process

Projects may have a substantial impact on the distribution of resources and status within the organization, with some departments/groups gaining and others losing, as a result of the project. Quite naturally, this creates political tension between the groups concerned and between the groups and the project team. Each group may attempt to push the project in a direction that suits the group's interests, even if it means radically changing or abandoning the project. There may also be tension between different project teams. This highly charged political environment is quite common, and it clearly hinders project management. Political manoeuvres are likely to constrain communication and co-ordination, and to add an extra, rather explosive element to project re-evaluation and the revision of plans and estimates. A project that is running late or over-budget may be easy prey for groups that stand to lose significantly if the project is completed.

Changing and Conflicting Objectives

In some cases, the original objectives of a project may be fairly loose, and in other cases, although the objectives were clear, major changes in the business, technological or organizational environment will force a change in the objectives. Other changes result from learning and politics processes discussed above. These changes may have a major impact on the management and planning of the project. Changes to the plan may also be required, when, for example, the original estimate was always unrealistic, the plan was incomplete, or major problems were encountered. The project manager needs to be aware of the consequences of such changes in order to maintain control of the project. Care has to be taken regarding: who justified the changes? When can the specifications be finally frozen?

Communication and Co-ordination

Communication and co-ordination between team members and between the team and the rest of the organization is a perennial problem, especially in large projects. Such projects may involve hundreds of specialists over a number of years. Small teams are much easier to manage than large 'armies', but sometimes the demands of the project necessitate a large work force. No matter how large the team, there is a need to develop common concepts and representations (of plans, reports etc.). Whereas good documentation is essential from this communication, we should not forget the importance of informal communication in the shape of emissaries and 'boundary spanners', who maintain informal links between groups.

Project Management is People Management

Above all, project management is people management. Usually the members of the project team are the most valuable resource of the project. They are the ones that are likely to make the difference between the project succeeding or failing. Thus considerable care should be taken in the recruitment, training and motivation of team members. Managers must take care to balance the needs of the team against those of the individuals. Each team member should feel that they are a key of the project team and not just an anonymous 'foot soldier'. When the pressure increases, the team should feel that this is a challenge, not just more exploitation.

Conclusion

The technical and technological aspects of project management discussed in this article may suggest that project management seeks to impose a highly formal, reductionist strategy on development. Yet we know that much project work is characterized as a process of learning, communication and negotiation. The appropriateness of formal project management can be seen by considering two very different scenarios. Firstly, for straightforward projects, i.e. projects that are highly structured, well-defined, of low risk, involving familiar technology and well understood techniques, the techniques of project management provide a strong framework for accurate planning, monitoring and control. However, the situation is not so straightforward in the second case of poorly understood projects, i.e. projects that are unstructured, badly defined, high risk, involving novel technology or processes. In this case, project management techniques may turn from enabling mechanisms into obstacles to the project's progress. Here, plans and estimates need to be interpreted sensitively and subjectively, and not treated as rigid, deterministic schemes.

It should be noted that the techniques presented here for project management are appropriate for large(ish), relatively complex projects, involving multiple resources; they are not suitable for maintaining 'to-do' lists. Project management takes a lot of (your) time and effort, but it can help to save projects. However, project management alone cannot transform bad projects/managers into good ones.

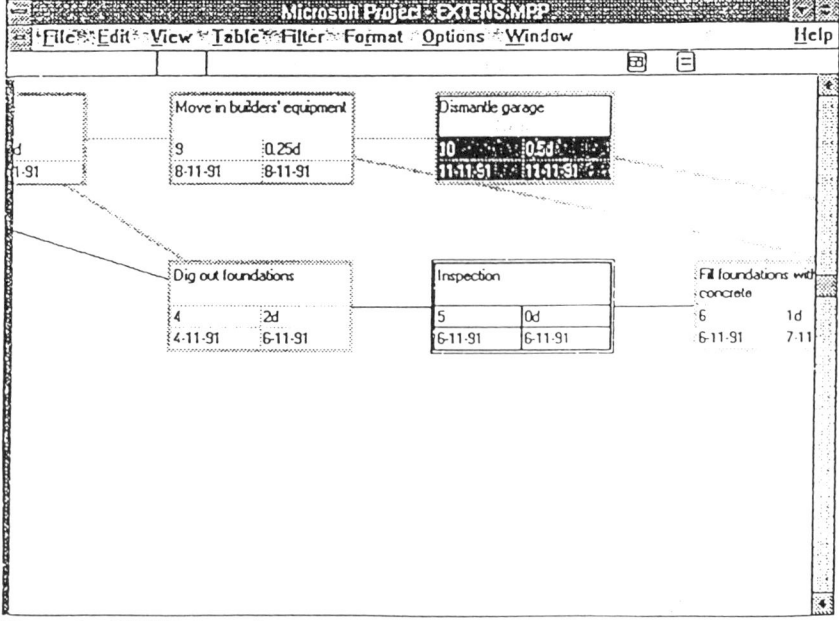

Figure 2 Gantt chart view

Figure 3 PERT chart view

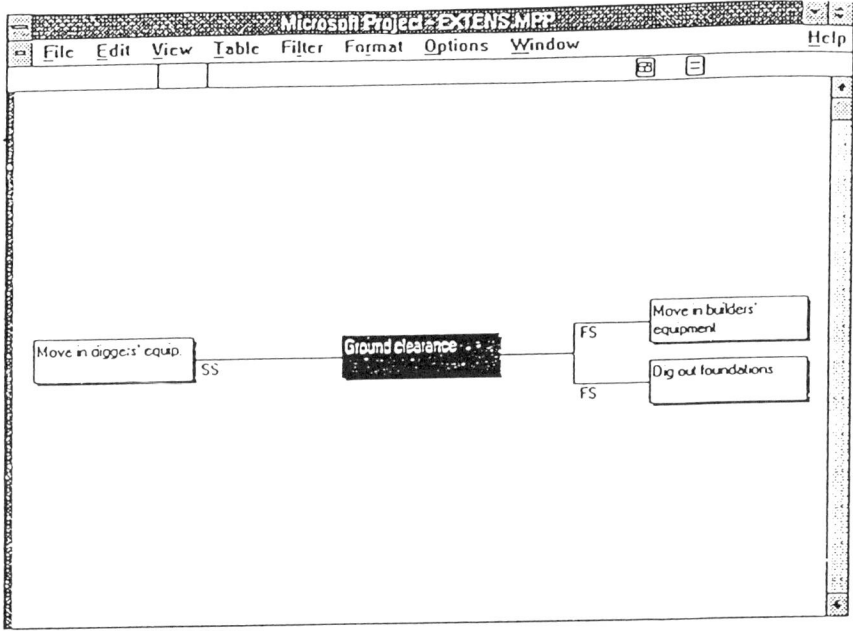

Figure 4   Task PERT view

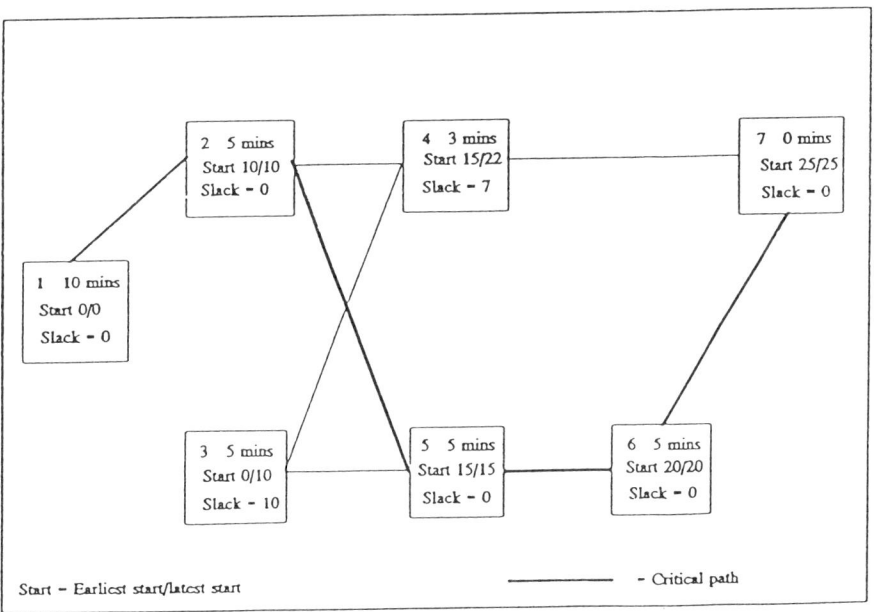

Start − Earliest start/latest start   ——— − Critical path

Figure 5   PERT Chart/Critical path method

Recommended Reading

Project Management
Angell, I.O. & Smithson, S. (1991), *Information Systems Management: Opportunities and Risks*, Chap. 9, Macmillan, Basingstoke.
Bergen, S.A. (1986), *Project Management*, Blackwell, Oxford.
Berkley, D., de Hoog R., Humphreys, P. (1990), *Software Development Project Management*, Ellis Horwood, London.
Brill, A.E. (ed) (1984), *Techniques of EDP Project Management*, Yourdon Press, New York.
Charette, R. (1989), *Software Engineering Risk Analysis and Management*, McGraw-Hill, New York.
Cooper, D.F. & Chapman, C.B. (1987), *Risk Analysis for Large Projects: Models, Methods and Cases*, Wiley, Chichester.
Gildersleeve, T.R. (1985), *Data Processing Project Management*, 2nd edn, Van Nostrand Reinhold, New York.
Kharbanda, O. (1991), *Project Teams: The Human Factor*, NCC Blackwell, Oxford.
Moder, J.J. & Phillips, C.R. (1970), *Project Management with CPM and PERT*, 2nd edn, Van Nostrand, New York.
Morris, W.G. & Hough G.H. (1987), *The Anatomy of Major Projectsd: A Study of the Reality of Project Management*, Wiley, Chichester.
Taffler, R. (1979), *Using Operational Research*, Prentice Hall, London.
Thamhain, H.J., Wilemon, D.L. (1986), *Criteria for Controlling Projects According to Plan*, Project Management Journal, June, 75-81.

Project Management Software
Wood, L. (1988), *Product focus: The promise of project management*, Byte, Vol. 13, No. 12, 180-192.
Informatics, Vol. 11, No. 1, January 1990, 48-64.
PC Magazine, Vol. 3, No. 9, September 1990, 106-128.

Spreadsheet Module

Edgar A. Whitley, Department of Information Systems, London School of Economics and Political Science, London, United Kingdom

Introduction

One of the most important factors behind the widespread use of personal computers in the management of organizations is the electronic spreadsheet. Spreadsheets are packages that provide managers with many of the facilities they require to analyze complex numerical situations in order to assess different alternatives, plan for the future and manage their organizations. A spreadsheet package is a very flexible modelling tool which, nowadays, also includes facilities for the graphical presentation of data as well as some database facilities.

A **spreadsheet model** is like a large sheet of paper where figures and relationships are entered. Because the model is built on a computer it is very easy to make changes in one of the figures and the effects of those changes are immediately seen in related areas of the spreadsheet. For example, by changing the number of goods produced related figures such as total cost of production, revenue and profitability will automatically be adjusted as well. The ease with which these changes can be made means that users can introduce variations in the model and examine the implications of the different scenarios. This process, which is often known as **what-if** analysis, allows for a better understanding of the problem situation and its solution.

Lotus 123 is one of the most successful spreadsheet packages available on the market and it will be used as the basis of this tutorial. Most of the principles that are described below, however, are equally applicable to other spreadsheet packages.

36

Figure 1   Basic Concepts

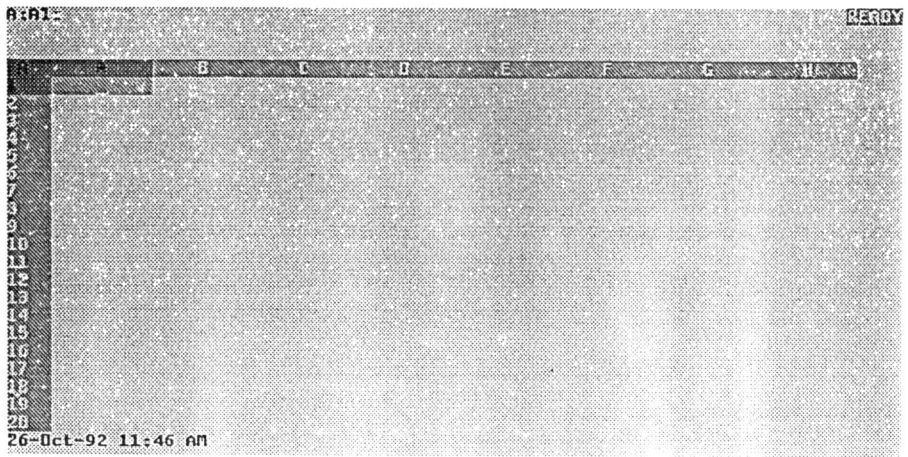

Basic Concepts

A **spreadsheet model** is like a large sheet of paper broken down into **rows** and **columns**. The rows are labelled with numbers and the columns with letters. Thus the top-left hand **cell** in the spreadsheet has the **cell label** *A1*.

Each cell in the spreadsheet can contain one of the following elements: a number, some text or a formula relating other elements together. To put an element into a particular cell, you must first move the **highlight** to that cell. To do this use the arrow (**cursor**) keys.

Figure 2

When it comes to entering data into a cell, the software tries to be intelligent about what kind of an element you are entering. Thus if you type something starting with a number it assumes that you are entering a number. The number that you type in appears at the top left hand corner of the screen, next to a label describing the highlighted cell. When you press Return the number you typed in the **input line** is displayed in the highlighted cell.

Figure 3

If you enter some data starting with a letter then the software assumes you are entering some text. This is often a **text label** used to describe the model and make it easier to understand than simply using cell labels. Any text that you enter is marked with a ' to differentiate it from other kinds of elements.

Note that once you have entered an element it is displayed in the input line at the top left hand corner of the screen. As you move around from cell to cell the contents of each cell is displayed at that point.

Whenever you type in a value it is put in the highlighted cell. If you want to change the contents of a cell that has already been filled, move the highlight to that cell and press F2. The contents will then be displayed in the input line and you can alter the existing value. When you then press Return the edited contents are placed in the cell.

Figure 4   Commands in Lotus 123

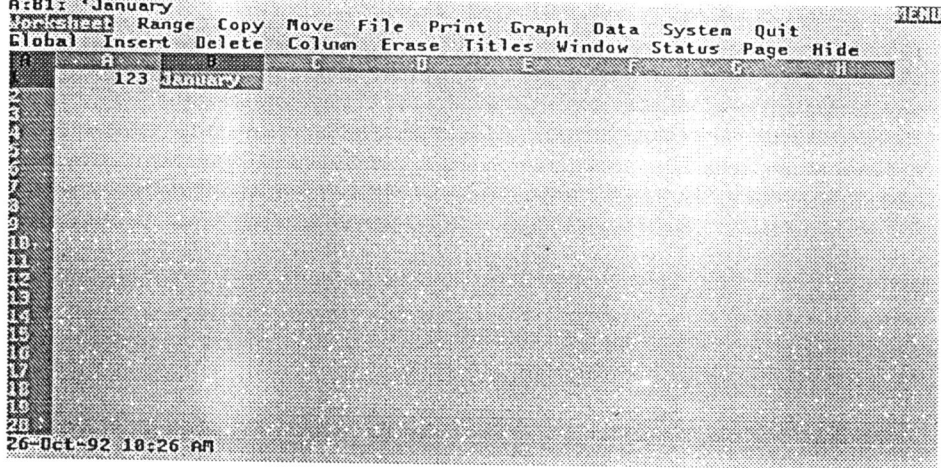

Commands in Lotus 123

In addition to entering data and formulae, Lotus 123 also provides a large number of commands. These are all accessed by moving through a series of **menus** to define the command that you want to execute. The menu structure is in the form of a 'tree' of sub menus in that the first menu brings up the main topic areas, and you can then move into those areas to have a more detailed **sub menu**.

To bring up the first menu press / (forward slash). This brings up the menu as shown below. The commands are built up in a hierarchical manner with the top level of commands describing the main functions that can be provided by the system. For example, there are commands that affect the whole worksheet, commands to deal with files and commands for data manipulation. You can select the command that you want in one of two ways. Firstly you can type the first letter of the command that you want to use - F for file commands for example. Alternatively you can move the highlight over the command required by using the cursor keys. On the line below the menu you can see the next level of commands available under that particular heading (for example, under worksheet the next commands are Global, Insert, Delete etc.) and as you move the highlight the available options change. Once you have moved the highlight to the required item press Return.

This process continues until you activate the chosen command (or need to specify data for it) or it moves you to the next level of the menu structure. Depending on the chosen command this can be a few or many keystrokes.

To exit from a particular sub menu, either select the Quit menu option (if it is available) or press Escape.

Spreadsheet Models

The spreadsheet models we have looked at so far have been very simple. Indeed they have been nothing more than a grid which allows us to enter text and numbers. Whilst in a business or planning situation it is important to be able to view the available data in a clear and structured manner, spreadsheets would not have been as successful and important to the day-to-day running of so many organizations if that was all that they could do. What makes spreadsheets useful is the way that they support the manipulation of the data and how they handle it in an automatic way. For example, in addition to listing a series of costs associated with a product in a particular month it is also possible to perform calculations on them - for example to calculate the total costs in a particular month.
This (calculated) result can then be combined with another calculated result (say the net revenue) to provide a third calculated result the profit. The automatic nature of spreadsheets means that if one of the costs changes then *all* the related values change immediately - thus total costs and profit will be adjusted to reflect immediately this change.

Furthermore, the spreadsheet allows for the display of this data in graphical form for presentation and inclusion in reports. The data in the spreadsheet can also be used in the form of a database. All these functions will now be described.

Operations on Data

The basic mathematical operations of addition, subtraction, multiplication and division can be performed on the numeric data in the spreadsheet. Addition is specified using +, subtraction using -, multiplication with * and division with / (forward slash). If you want cell A3 to contain the sum of cells A1 and A2 then you enter a *formula* like +A1+A2 when the highlight is positioned over cell A3.

Note that the editing line at the top of the screen displays the formula whilst the cell on the grid displays the result of the calculation. It was necessary to prefix a + to the formula because otherwise the system would

40

think that you were entering a text label starting with the letters A1. The use of a + ensures that the input is considered as a formula.

Whenever the contents of cell A1 or A2 changes the result of the calculation in cell A3 also changes automatically.

The formulae used can be more complicated - for example, you may want to divide the sum of cells A1 and A2 by the difference between cells B3 and B4. The formula would be +(A1+A2)/(B4-B3). Note how brackets are used to ensure that the calculation is performed

Figure 5

in the correct order as +A1+A2/B4-B3 would be interpreted as dividing A2 by B4, adding A1 and then subtracting B3.

Figure 6

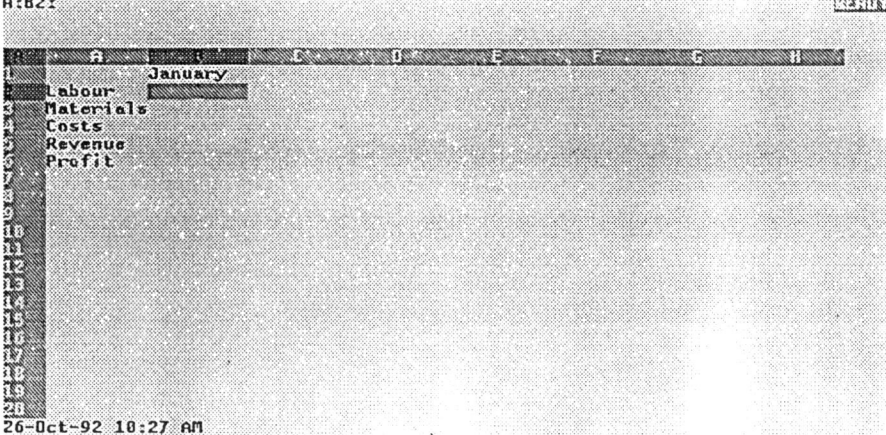

To demonstrate these principles, let us develop a simple model of a company which sells a product. In the first column we will have text labels which describe the various parts of the model - labour, materials, total costs, revenue and profit. The first row contains text labels describing the different months.

We can now define the formulae that will enable our model to automatically provide the profit for the month. We begin by defining the total cost in January. This is given by adding the labour B2 to the materials B3 and storing the result in cell B4. The formula for cell B4 is therefore +B2+B3.

Figure 7

A:B6: +B5-B4   READY

	January	
Labour	100	
Materials	50	
Costs	150	
Revenue	200	
Profit		

26-Oct-92 10:28 AM

The profit is calculated as revenue B5 minus the total costs B4. The formula for cell B6 is therefore +B5-B4.

Saving

Figure 8

```
A:B6: +B5-B4                                                      MENU
Retrieve  Save  Combine  Xtract  Erase  List  Import  Dir  New  Open  Admin
Replace the current file with a file from disk
        A               B           C       D       E       F       G       H
                 January
1
2    Labour          100
3    Materials        50
4    Costs           150
5    Revenue         200
6    Profit           50
7
8
9
10
11
12
13
14
15
16
17
18
19
20
26-Oct-92 10:28 AM
```

Now that we have developed a small model it is probably a good idea to save it on disk so that we have a permanent record of it. To save the spreadsheet model on the disk we need to first call up the main menu. As was described earlier, this is done by pressing the / (forward slash) key. This brings up the menu and we want to select the File sub menu. This can be done by either moving the highlight to File and pressing Return or by pressing F. We are now faced with the File sub menu and we want to Save the spreadsheet so we press S. The system now asks us to specify the name of the model and so we enter SIMP-MOD and press Return. The system has now saved the spreadsheet on disk and given it the name SIMP-MOD.WK3. The .WK? extension is used by Lotus 123 to signify a spreadsheet file and .WK3 signifies that it was produced by version 3 of the package. The filename is now shown at the bottom left hand corner of the screen.

Figure 9

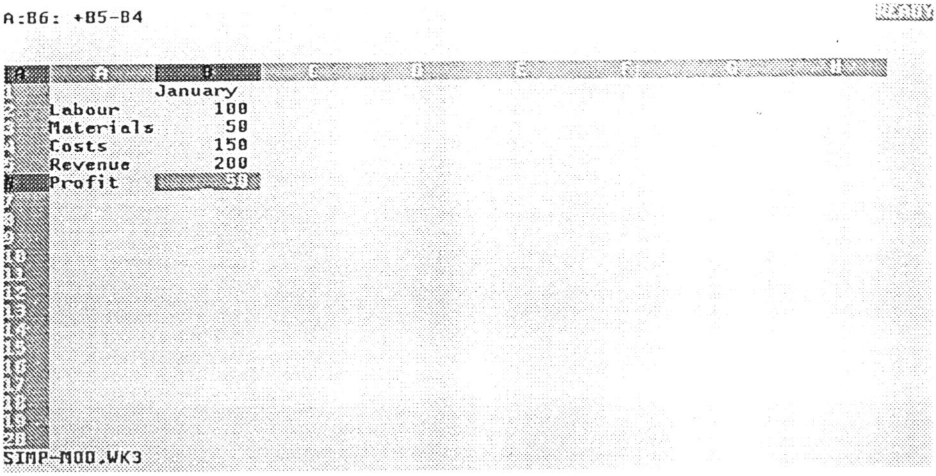

```
A:B6:  +B5-B4                                                    READY
```

```
              January
   Labour         100
   Materials       50
   Costs          150
   Revenue        200
   Profit          50
```

```
SIMP-MOD.WK3
```

Clearing the Spreadsheet

To clear the spreadsheet that we have been working with enter the following command / Worksheet Erase (/WE). This deletes the current worksheet from memory and is a permanent step. If you have not saved the current spreadsheet to disk the package will ask you to confirm that this is what you want to do since the action is irreversible. After entering this command you are faced with a clear spreadsheet.

Loading a Spreadsheet

To load a spreadsheet from disk you enter / File Load (/FL) and then specify the name of the spreadsheet you want (for example, SIMP-MOD.WK3). The system will display the available spreadsheet files on the screen and you can select the required one by using the cursor keys and pressing Return.

Ranges

Our spreadsheet model at the moment only contains one month's data and it would be nice to be able to extend it, for example, to the first six months of the year. This could be done in a number of ways. The most time consuming would be to enter data for each month and to re-enter the formulae in each column.

Fortunately, the package allows you to do this in a far easier manner by copying a **range** of data. Before describing how ranges are copied we will first examine the notion of a range in a spreadsheet.

Figure 10

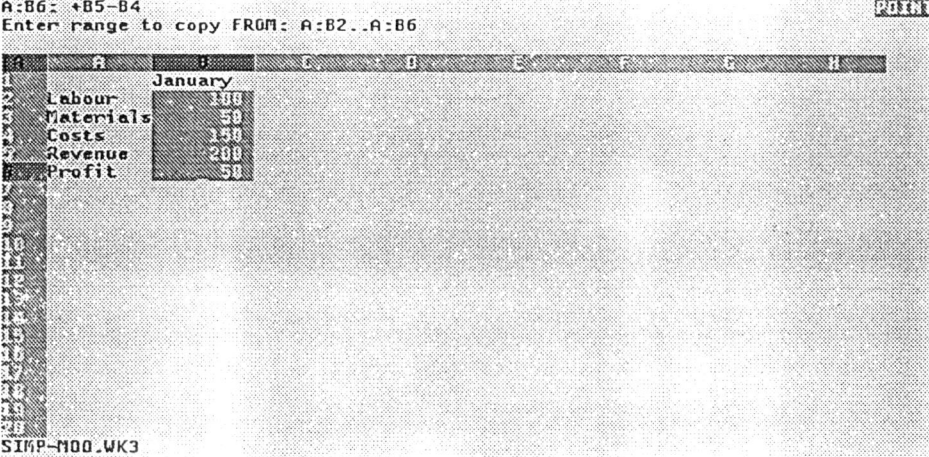

Figure 11

A range is simply a rectangular area in the spreadsheet model. It is defined by giving the location of the top left hand corner of the rectangle and the bottom right hand corner of the rectangle, separated by two full stops. Thus a range that extends from B2 to C6 would be labelled as B2..C6. This would include cells B2, B3, B4, B5, B6, C2, C3, C4, C5 and C6 and corresponds to the shaded area in the diagram below.
Note that although ranges are normally rectangles it is possible for a range to refer to a single cell. Thus the range C7..C7 is identical to the cell C7.

When defining ranges in spreadsheet commands two methods can be used. One approach is to simply enter the range at the keyboard. However, this requires that you know exactly what the range is. If this is not the case then you can *mark* the range on the grid. When you are required to specify a range you can move the highlight to one of the two corners of the grid and then press the full stop key. This now allows you to move the highlight to the opposite corner and all the range in between will be highlighted. If the range is starting at the wrong point then press Escape. The system also automatically updates the range specification as you move the highlight around.

To copy a range you call up the Copy command. This is done by entering / Copy (/C). You are then asked to specify the range to be copied. Using either of the methods described above select the range B2..B6. After pressing Return you must move the highlight to where you want the range to be copied to. Move the highlight to cell C2. If you press Return now the range will be copied to column C. If you want the range to be copied to a number of columns then simply select a range of columns (e.g. C2..G2) and the data will be copied to each of the columns. Note that if you copy to an area of the spreadsheet model where there is already data or formulae then the existing values will be overwritten by the newly copied values. All that now needs to be done is for the labels to be added for the remaining months. By using the move command / Move (/M) you can move rather than copy a range of data.

Relative Markings

When the formulae were entered in column B they referred to cells in the same column. The copied formulae in column C, however, refer to cells in column C rather than column B. This is a feature of spreadsheets which considerably simplifies the development of large models.

All the references that we have specified, for example +B2+B3 in cell B4, do not actually refer to cells B2 and B3. What they refer to are the cells that are one and two positions above the current cell. Thus, for example, if the formula is copied to cell G56 it again refers to the cells that are one and two above the current one, i.e. G54 and G55.

If you want to refer to an **absolute cell position** (rather than a **relative cell position**) then you must use the $ symbol in the reference. If you use a $ before the column then the column is held absolute, a $ before the row holds the row absolute. Therefore to refer to a particular cell you must use $ before both the row *and* the column.

We can use this notion of relative referencing to develop our model further. Let us assume that in the second month the labour costs increase by 5%, materials increase by 1% and revenue increases by 2%. Instead of entering data for the second month's costs and revenue we can enter formulae. Thus in cell C2 (February's labour) we have January's labour multiplied by 1.05 (for the 5% increase). The formula is thus +B2*1.05. Similarly, in C3 we have +B3*1.01 and in C5 we have +B5*1.02. The addition in C4 is +C2+C3 and in C6 is +C5-C4.

If we wanted our model to be more flexible we would not use **constants** in these formulae, instead we would store our multiplication factors in a range of cells and would refer to them instead. Thus if cell B10 contained the labour cost increase multiplication factor then the formula in C2 would be +B2*B10. If we wanted the same percentage increase over all the months then this would have to be an absolute reference +B2*B10.

By making use of these relative cell position references we can copy these formulae so that labour costs rise by 5% each month based on the previous month. By using the copy command the formulae are automatically updated to reflect their new positions.

Functions

Although many common operations can be performed by simply using the arithmetic
operations described previously, Lotus 123 also provides a large number of functions which can considerably enhance the functionality of the model. All the functions are prefixed with the @ symbol. One of the most widely used functions is the @SUM function which provides a sum of the values in the range specified. Thus to add the contents of the cells A2, A3, A4, A5, A6, A7, A8, A9 and A10 you can either enter the formula +A2+A3+A4+A5+A6+A7+A8+A9+A10 or you can use the more convenient function @SUM(A2..A10). A list of the most commonly used functions is provided at the end of this document.

Formatting Data

The data that is presented in the spreadsheet can represent many different kinds of numbers -numeric results, approximations, scientific results, currency, percentages etc. It is possible to specify how the spreadsheet should present these different kinds of numbers by making use of the formatting commands available in the system.

Formats can be applied to either ranges - if you want to change the appearance of a particular set of data - or to the whole spreadsheet - for example, if all the numbers are currency.

If we want to display some of the data in the spreadsheet as currency the following steps would be taken. We select the Range command from the main menu and then the Format command (/RF). We then specify that we want currency (C) and the system asks us how many decimal places we want - the default response is 2. We are then asked to specify the range that is to be formatted in this way. Once this has been done all the figures in the range are displayed using the default currency display format. This would be of the form £100.00. If you want to change the default currency display format then enter / Worksheet Global Default Other International Currency (/WGDOIC). You are then asked to enter the currency symbol and then specify whether it comes before the value or after it. Ther select Quit twice to return to the spreadsheet where your changes will be shown.

If you want all the figures in the spreadsheet to be displayed as currency then enter / Worksheet Global Format Currency (/WGFC) and specify the number of decimal places. All figures that are entered *after* this command will be shown in the new format. To change the format of existing values you must use the range format command.

Other formats that are commonly used are fixed (for example, show all the figures to three decimal places), scientific (e.g. 1.2E2), percent and hidden (which hides the contents of the cell).

When using the format command you may find that the system displays the contents of a cell as a series of asterisks. This signifies that the current cell width is not enough to display the contents properly and the system is therefore not displaying them at all. When this arises it is important to set the width of the column so that the data can be viewed. To do this select / Worksheet Column Set-width (/WCS). You are then asked to enter the new column width. This can be done in two ways - either by entering the value at the prompt or by pressing the left or right arrows to change the column width.

Other related useful commands include the ability to fill a cell with a particular character regardless of the width of the cell. This is done by entering \ (backward slash) followed by the character to be entered. Thus entering \= will cause the cell to display ======= to its full width. If the width of the column is altered then the display will be altered accordingly. To insert new rows into the spreadsheet enter / Worksheet Insert Row (/WIR) and then specifying the range of rows that are to be entered. To insert columns the / Worksheet Insert Column (/WIC) command is used and there is a related command for deleting rows and columns.

Printing

Another useful command is the ability to print the data in the spreadsheet. Printing involves a number of stages. To print to a printer (rather than indirectly to a file) you enter / Print Printer (/PP). You are then required to specify the range to be printed (R). This is done as described above by specifying the top-left and bottom-right hand corner of the range to be printed. Next issue the Align (A) command which will automatically ensure that the printer is at the top of the page and then press Go (G). The printing will then start.

Displaying Data Graphically

Spreadsheets also offer the ability to display the numeric data in a variety of graphical forms, many of which can become very sophisticated. It is beyond the scope of this introduction to describe the full range of graphics available, but the main principles will be outlined.

In order to create a graph you need a set of data and a chosen graph type. You then specify how the data you want to display relates to the graph type. For example, let us consider the monthly profit figures that we have created in our model. One way to display this would be as a monthly profit figure in a line graph so that the overall trend can be seen.

The command / Graph (/G) takes you to the graph sub-menu and the first thing to select is the graph type. Since we want a line graph this is Type Line (TL). All the graphs make use of a number of different data ranges, labelled, X, A, B, C, D, E and F. The X data range is normally used to describe the labels for the x-axis of the graph with the A to F data ranges being used to specify the data to be displayed. In some cases, such as pie charts described below, this is not the case.

For the line graph the A data range contains the values to be graphed so enter A and the range of values to be graphed (A B6..G6). The graph can now be viewed (V). As it stands, however, the graph does not tell us very much. The next thing we want is an indication of the time scale involved. This is done by specifying the X range to be the text labels that we want (X B1..G1).

Figure 12

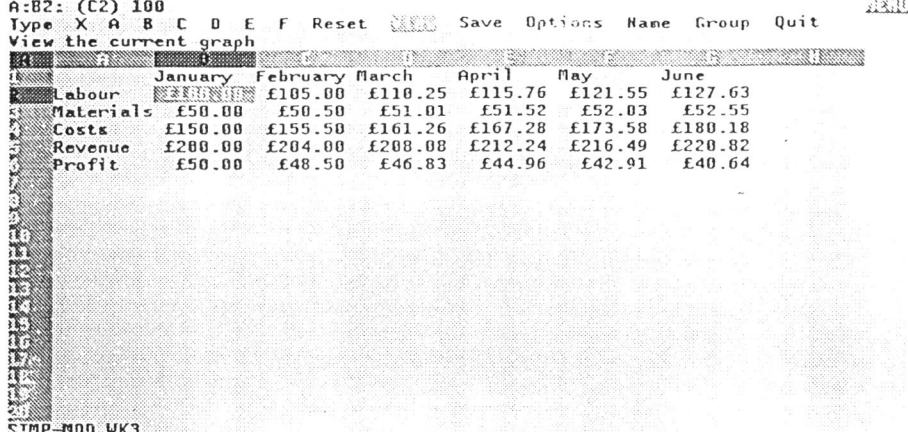

To add a title to the graph use Options Titles First (OTF) and then specify the first line of the title. Titles Second (TS) specifies the second line of the titles. If you want to view other data on the graph then mark them as ranges B through G. If you want a legend to describe the different lines then enter Options Legend A and specify the name of the first range. If you enter \ you can then refer to a cell in the spreadsheet for this information.

Figure 13

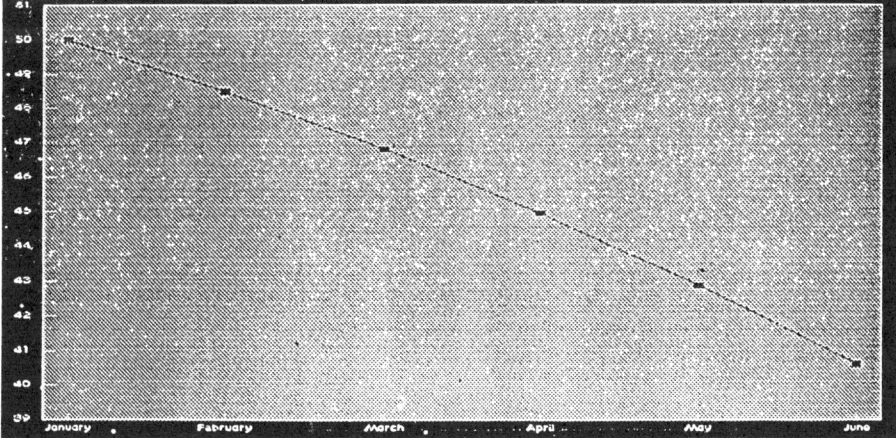

The last important feature is the idea of naming a graph. Since you may want to have more than one graph associated with any particular spreadsheet it is a good idea to name each graph that is required. Naming the graph saves all the information about the data ranges used to display it etc. To name a graph enter Name Create (NC) and the graph's name, perhaps Graph1. When you want to reuse the graph enter Name Use (NU) and select the Graph1 graph from the list of available graphs.

If you want to view the breakdown of costs in January then a pie-chart is probably the best display. To create a pie-chart select / Graph Type Pie (/GTP). The A range is used to specify the data to be displayed by the pie chart (A B2..B3). The X range is used to specify the pie-slice labels and the B range is used to specify the colours of the different slices. If the slice colour has a value greater than 100 then that slice is 'exploded' from the pie-chart.

Database Type Operations

Figure 14

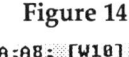

Spreadsheets like Lotus 123 also offer the ability to perform simple database type operations on the data in the spreadsheet. To demonstrate this facility, let us construct a simple example of lecturers who teach different courses. For our database, let us have three fields -Name, Title and Course Taught. We begin by entering text labels in a row for the three fields in the spreadsheet and then enter some data below them.

Next we copy these text labels into two further parts of the spreadsheet model. The original part contains the **input data** we are working on, the second contains the **criteria** we will select on and the third will be used for displaying the **output results** of the query.

To query the data we move to the second part of the model (the criteria) and enter the values that we are looking for. Suppose we want to select all those people who have the title Dr. To do this we enter Dr under the Title text label in the criteria part.

Now we need to access the data related commands. This is done by first defining the three ranges. / Data Query (/DQ) and then specifying the input (data) range (I). Note that the text labels we defined must be included in the range as they tell the spreadsheet where to search for the data. Next specify the criteria range (C). Again include the labels. Finally, define the output range (O). Querying the database can be done in two ways. One way is to have the system highlight the matching input records. Alternatively the matching records can be extracted to the spreadsheet in the output range. To highlight the data use the Find (F) command and the highlight will be placed over the first matching record. To move to the next matching record press the cursor keys. To finish the highlighting process press Escape. If you want to extract the data enter the Extract command (E) and the matching records are copied to the output range.

Figure 15

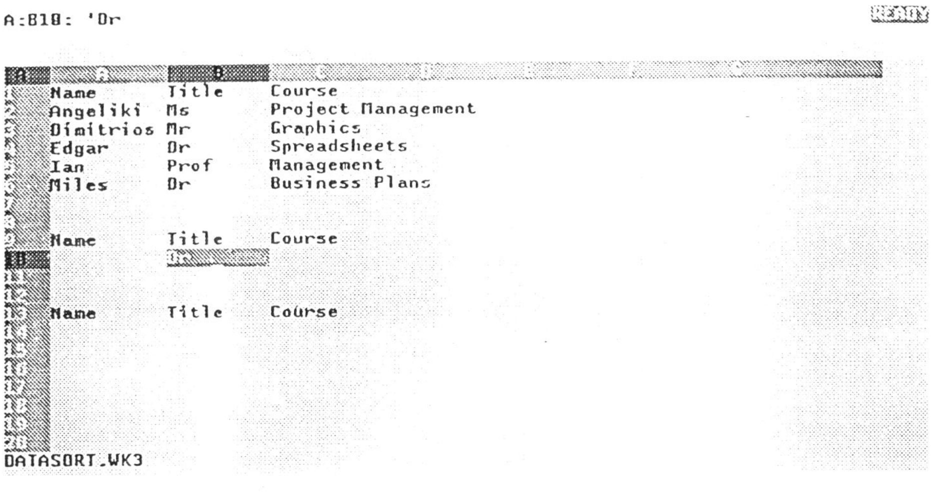

A:B18:   'Dr   READY

Name	Title	Course
Angeliki	Ms	Project Management
Dimitrios	Mr	Graphics
Edgar	Dr	Spreadsheets
Ian	Prof	Management
Miles	Dr	Business Plans
Name	Title	Course
Name	Title	Course

DATASORT.WK3

The system operates a form of Query-By-Example in that it tries to match the example in the criteria range with the data found. In the illustration above, the example is a value for the Title field. Any data which matches this example in the Title field is selected by the system. The match can be a direct match or a comparison (for example, if a field contained numbers you could specify that the value be >21). If the example covers more than one field then the match *must* be made on all parts of the example. It is possible to specify a number of examples in the criteria range and if any of them match then they are selected.

Data Sorting

The data in our database can also be automatically sorted by the system. In order to do this we must specify the data range, the key which is being used to sort the data and possibly the secondary key to further sort the data. / Data Sort (/DS) allows you to specify the Data-Range (D) to be sorted and the Primary-Key (P) which specifies the data label to be used to sort the data. You are also asked to specify if the primary key is to be sorted into ascending or descending order (A or D). If there are duplicate Primary-Keys then the Secondary-Key (S) can be used to further specify the sort. Again you specify if this is to be in ascending or descending order. To perform the sort enter Go (G) and the original data will be replaced by the sorted data.

Data Distribution

Data distribution is another useful command offered by Lotus 123. This is a means of counting how many data values fall within certain bounds and is particularly useful for examination mark type data. Suppose you want to know how many of the marks exist between 10 and 20, between 20 and 30 etc. The command needs two sets of data. The first is the data that is to be classified, the second is the distribution (called the Bin range) that you wish to examine the spread of data over.

/ Data Distribution (/DD) asks you to specify the values range followed by the Bin range. The results are placed in the column immediately to the right of the Bin range. The first result is the number of values less than or equal to the first value in the Bin range, the second result is the number of values greater than the first value in the Bin range and less than or equal to the second value in the Bin range etc. Note that the data does not need to be sorted for this command to work.

Useful Functions

@ABS(X)	Calculate the positive (absolute) value of X
@AVG(Range)	Calculate the average value in the specified range
@COUNT(Range)	Count the number of non-blank entries in the range
@INT(X)	Calculate the integer portion of X without rounding the value.
@MAX(Range)	Find the largest value in the range
@MIN(Range)	Find the smallest value in the range
@PI	Return the value of π
@SQRT(X)	Calculate the square root of X
@STD(Range)	Calculate the population standard deviation of the range
@STDS(Range)	Calculate the sample standard deviation of the range
@SUM(Range)	Calculate the sum of the values in the range
@VAR(Range)	Calculate the population variance of the range
@VARS(Range)	Calculate the sample variance of the values in the range.

Cost Management for New Technology

Miles B. Gietzmann, Department of Accounting and Finance, London School of Economics and Political Science, London, United Kingdom

It is of central importance for the individual nations of the former Eastern block countries and the developing world nations to introduce technologically advanced manufacturing processes, in order to be able to compete in the new world order. This requires the application of the latest scientific (engineering) discoveries and practices. The establishment of such scientific knowledge bases is an essential prerequisite. However in of itself it is not sufficient.

There are many choices to be made with respect to technology and resulting product specification. That is the scientific knowledge needs to be applied in a way that results in products actively demanded by consumers. This in turn leads to effective wealth creation for successful manufacturing organizations which can be reinvested, for instance in enhanced productive facilities and worker training. A phrase commonly used in the West to describe the process of evaluating the potential introduction of new technology is "strategic management". One is required to relate the costs of introducing new technology, to the potential benefits. This is difficult because in a competitive environment the organization must try to understand the complex links between competitor reactions and consumer taste variations. The organization needs to critically evaluate what "value" it is passing on to consumers when they purchase its product and why or if other firms cannot replicate the value provision service.

The first step in such a process is to build up best guess forecasts. However, given the uncertainty attached to these forecasts one needs to critically evaluate how sensitive, forecasted financial performance is to assumptions (forecasts) about key variables.

In order to introduce participants to the principles of cost management of new technologies I will make use of the Lotus 123 spreadsheet package introduced in the previous section. I will introduce a stylized organization called the House of Mark which serves as a practical example for utilizing the techniques outlined previously. For example, formulae will be entered into the spreadsheet which represent the income received from selling goods to customers who pay cash and receive a discount for paying by cash.

The spreadsheet model that is developed will be used to demonstrate the difference between two forms of cost management, namely cash flow accounting versus accrual accounting. Cash flow accounting recognizes a transaction when the cash receipt or payment takes place. Accrual accounting, in contrast, recognizes revenues when goods are sold or when services are rendered. These ideas are then built into the spreadsheet model Lotus 123.

Analysing the cash flow budget that has been developed using traditional ordered accounting logic is rather difficult and I introduce a more structured design process, called exogenous/endogenous formulated spreadsheets. In these spreadsheets, the model is broken into two components. The exogenous variables are the determiners in the model, such as required closing inventory. Endogenous variables are the results of formulae in the model. By separating these two forms of variables, it is far easier to systematically explore the model and examine the assumptions that underlie it.

Budgeting is concerned with using forecasts of uncertain exogenous variables to predict the future of key decision (endogenous) variables. If changing an exogenous variable's value does not significantly change the budget result then uncertainty with regard to this variable is not likely to be of central importance and vice versa.

In order to systematically determine importance one must parameterize a problem so that every endogenous relationship is entered in formula theoretic style. This means that all the endogenous variables do not have any explicit values, only references to cells containing the values. Thus instead of having a formula containing a 50% markup on the cost price, the formula would refer to the cell containing the markup value (as an exogenous variable).

The budgeting process can be automated by making use of Lotus 123's Data Table facility. This allows the systematic evaluation of a number of values for an exogenous variable to see the effect on a selected endogenous variable.

The results of the Data Table analysis can then be displayed in a variety of forms, using the graphing capabilities of Lotus 123 to provide an instant visualisation of the model.

The sensitivity analysis procedures developed in this part of the course could then be used in practice to evaluate investments in new technology. Thus what is being proposed is a need to subject potential scientific (engineering) innovations to careful financial scrutiny. The process of constructing such business plans should lead to an enhanced understanding of the financial implications of introducing new technology. This is required to ensure the continued existence of the organization and the establishment of financially competent technology development paths.

In order to be able to evaluate the financial performance (cash flow) of investments in new technology in a wider context, it is important to include the implications of the projected results, on traditional financial statements. The next component of the course introduces participants to traditional financial reporting statements. These are the Balance Sheet and the Income Statement (Profit and Loss Account). The importance of this exercise is recognized when one considers the role of investors. The principal financial statements by which they will appraise their existing investment or potential for new investment, are the Balance Sheet and Income Statement. Thus another important evaluating step in the technology appraisal process is an understanding of how such investment will impinge on these two key statements and the subsequent forming of investor expectations.

Lotus 123 provides an effective modelling tool to consider the inter relationships and facilitate the appraisal procedure. The limitations of such modelling processes should be clearly understood. For instance how realistic is it to assume one can model organizational specific investor expectations and reactions for the projected time span of the project? Again sensitivity analysis has a role to play in structuring expectations and understanding inter relationships only to the extent that they can be effectively modelled.

The final component of my course considers database manipulation. I demonstrate how Lotus 123 can be used as an effective database sorting and searching tool. Such techniques have widespread applicability. For instance searching a debtors ledger for individual accounts that meet certain criteria. In order to demonstrate the relationship between database commands and the sensitivity methodology discussed above, several problems are posed which require use of Lotus 123 database commands concurrently with Data Table commands.

Demand Analysis: International Market of Scientific Products

Sam Nilsson, Innovation Institute, Stockholm, Sweden

The Market for Scientific Products

A scientific product is defined as a product or process emerging from scientific institutions or industry.

The demand for scientific products may come from the scientific community itself and/or from industry and society at large.

We may distinguish between technology push or market pull. See Figure 1.

Market Pull

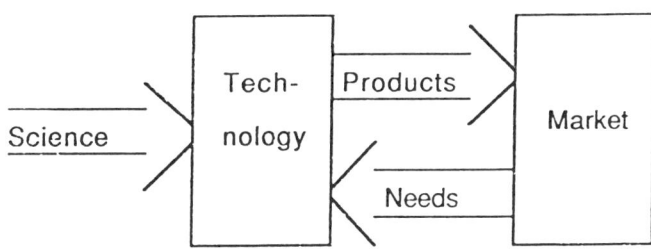

Technology Pull

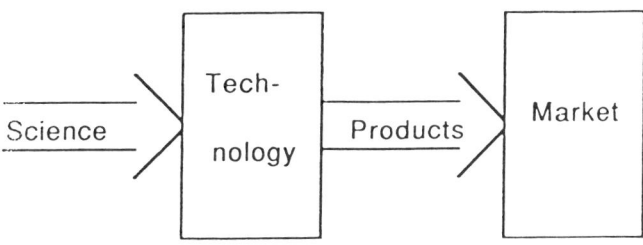

Figure 1

In many instances of new technology or scientific products there is no demand since the market is not aware of the possibilities of the new technology.

TV, when it was introduced and medical products emerging from contemporary academic and industrial research are examples of scientific/technological research which are not known by the market. In such cases the first demand must be created by "technology push".

However, as the market (society) becomes more and more sophisticated and demanding there will be increasing feed back from the market which in turn determines the next phases of the development of the scientific products.

A special case of needs response to scientific/technological development is that of developing countries, in which the market pull is often non-existent. At the same time it is evident that science and technology could have dramatic effects on the life conditions (fresh water, medical products, telecommunications, etc.) and economic development of these countries. In such cases help from outside is needed to build an infrastructure and intellectual capability so that the developing countries will be able to determine the appropriate scientific products for their development.

In the industrialized countries the signals from the market will be more and more decisive for the development of scientific products. Potential customers are involved in the specification of the new products together with R&D people. We are talking about "total customerization" which means highest possible quality and reliability. The so called intangibles (design, zero defect, reliability, service) are becoming more important than technological performance per se. Personal computers and cars are such cases. Traditional market research is a waste because it usually gives the solutions of "yesterday". In certain cases of very rapid scientific/technological development even planned obsolescence will be part of the product development.

Furthermore, the development of scientific products must consider the global market from the outset. This makes it even more necessary to work in close collaboration with the potential customers in order to reduce the risks of failures.

In conclusion:

> take time to decide what the market really needs;
> collaborate with customers as much as possible;
> consider the global market from the beginning;

"lead" the development rather than "react";
quality, reactability and service are more important than technical performance;
right timing is of crucial importance;
for developing countries local capabilities and infrastructure must be developed.

Product Cycles and Strategic Management of Product Development

Typical for product cycles is their logistic behaviour, i.e. they start from an idea or concept, grow exponentially on the market, mature and eventually die.

The most difficult strategic challenge is to bridge the discontinuity between the "old" product and the introduction of new product(s) to sustain the market position. See Figure 2. For companies or institutions to master discontinuities of product development it is necessary to follow trends both on the market and in science and technology. When 700 corporate leaders in Europe some years ago were asked what they perceived as the most difficult challenge during the 1990's the answer was "management of technology" Figure 2 encapsulates what they meant.

The transition from one product cycle to another is especially difficult because it is mostly not possible to extrapolate the technological characteristics of one product to a new generation of products because the transition depends on radical innovation rather than incremental innovation. Two examples: It would not have been possible to predict the development of Nylon from the properties and market position of Rayon or Cotton. Nylon was an entirely new product emerging from basic and applied chemical research. Similarly it would not have been possible to predict the transition from electric power production based on coal or oil to that of nuclear power, which emerged from basic atomic research in the 1930's and the discovery of heavy elements in the periodic system. In addition, the development of nuclear power for the civilian sector has been heavily subsidized by military development.

DISCONTINUITY

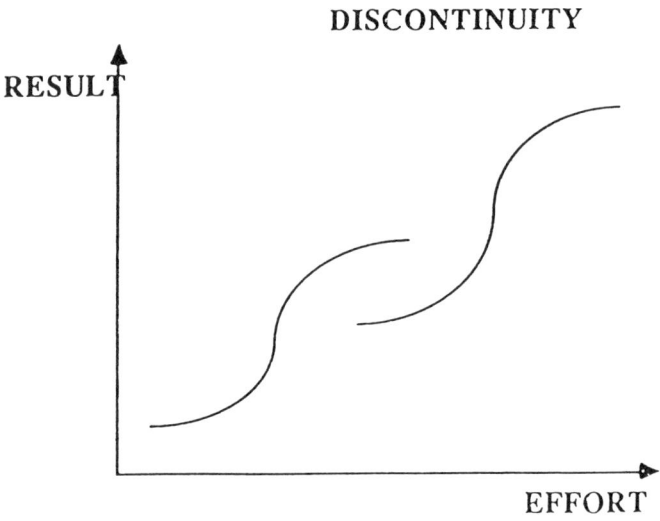

Figure 2

Below is given a list of warning signals for product cycles:

Warning signals indicating when one is reaching the end of a product life-cycle:

1. Consensus that present R&D results are unsatisfactory;
2. Increased number of missed deadlines and cost exceeding;
3. R&D focus shifts from product to process orientation;
4. Lowered creativity in R&D;
5. Low spirit and worries among R&D staff;
6. Potential business expansion based on narrow market segments;
7. Large differences in R&D costs between competitors without visible results;
8. Changes in R&D management without visible results;
9. Narrow market niches lost to small new business.

From a <u>strategic</u> development point of view there will never be a completely safe route to follow. A company or scientific/technological institution must always be prepared for surprises. However, if there is a sufficiently flexible and well informed management it is possible to counter the discontinuity surprise by adequate action. Below are given some recommendations regarding strategic management of product development.

Introduce hybrid products quickly;

Optimize costs and prolong the life of "old" products;
Create a vision of the future;
Communicate the vision at all levels of the organization;
Create a preparedness and will for change;
Be humble.

When it comes to the introduction of new products or new technology the following is worthwhile to keep in mind especially for a company.

Business sector within the organization responsible for new opportunities;
Committed person (entrepreneurial type) as driving force;
Support from top management necessary;
Co-operation with customer;
Timing very important ("Time-to-Market");
Encourage and awards (money or medals) to the "doers".

The development cycle of a new product is shown in Figure 3. Typically the investment phase is three to six years, if not longer. The Japanese companies are particularly good in making the "Time-to-Market" much shorter ($\approx$ 50%) than companies in Europe and USA. One reason for this is that the whole cycle from "basic" research to applied research and production is kept within one and the same company. In Europe and USA the new product development is based on co-operation with the universities or spin-offs from military research which tend to be less efficient because of its sequential nature (one step at a time).

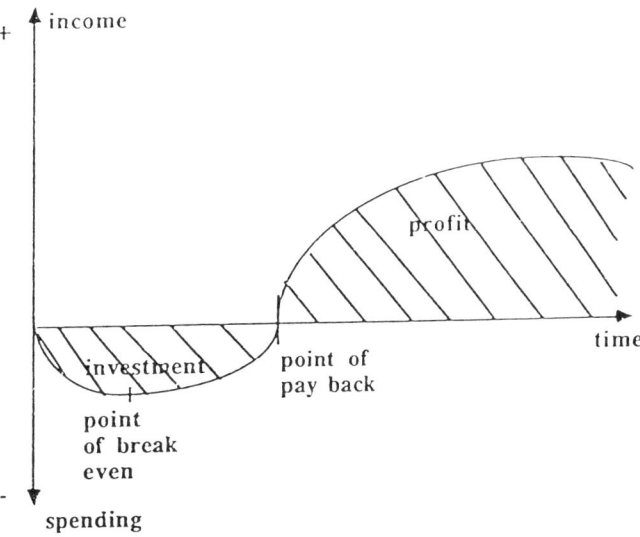

Figure 3

The basic dilemma is that it always seems to be more economical to protect the old products than to promote business development based on new products from research.

There are many different ways of keeping informed about long-term product development and new scientific products.
Among the intuitive methods the "Delphi" technique was quite popular a few years ago. It uses experts in a few rounds of independent and self-adjusting interviews, to make intuitive guesses about the time periods for scientific or technological breakthroughs. The use of independent consultants is another intuitive method which is more adaptable to the specific needs of the customer.

Another group of analysis of long-term product development is based on the notion of diffusion into the market of new products or technological concepts. Straightforward trend analysis gives very reliable predictions when a product is well established on the market, but it says nothing about new products.

Marchetti has developed a theory for the prediction of market penetration as well as innovation shifts of products, technical systems, such as energy sources, and even social phenomena. One example is shown in Figure 4.

However, the Marchetti method says little or nothing about scientific or technical breakthroughs.

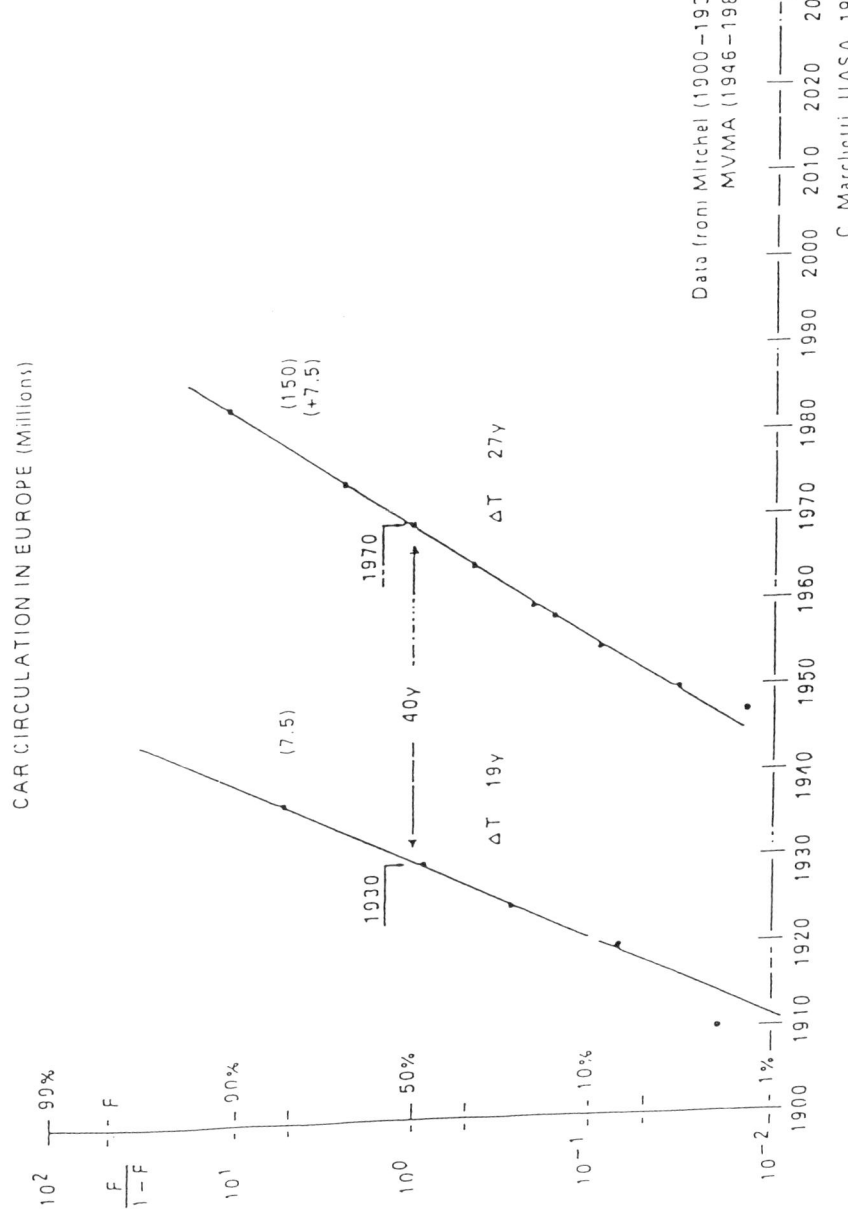

Figure 4

A third category of analysis of scientific products and market prospects is systematic evaluation of all relevant information wherever it comes from. The information can be anything from Nobel prizes to patents and scientific journals.

Information has become a new production factor. Strategic or Business "Intelligence" will be more important in the future for product development than systematic attempts to try to project into the future certain scientific or technological trends.

The capacity to collect, digest and use creatively the constant flow of information will be the most competitive advantage for any organization in the future. It is for this reason that "Strategic Management" will be more important than "Strategic Development". The real challenge is to be able to manage discontinuities and rapid change.

In this context it may be increasing to divide the different periods of management in Europe in the following way.

1945-60:	Production driven;
1960-74:	Marketing driven;
1974-82:	Finance driven;
1982-91:	Strategy driven;
1991-	Customer driven (high flexibility).

The Environment of Research, Development and Innovation

Research and innovation should not be carried out in a vacuum. They must recognize the major trends of social, political and economic development which ultimately affect them. Some trends are more predominant than others. The demographic development in the world suggests that the world population will increase from today's 5,600 million to about 8,500 million in year 2025.

Asia and Africa alone will contribute by 6,500 million at 2025. Another remarkable prediction is that out of this population of 8,500 million some 5,100 million will live in urban areas, and 75% of the total population in coastal areas. Already these figures indicate the enormous needs for infrastructure improvements especially in developing countries as well as improvements of health care and other social services. The needs for scientific, technical and social innovations and management of complex systems are staggering. Another trend which will affect the scientific and technical development is the rapid expansion of global trade. It is particularly important that the free trade systems (GATT) is expanded to include more and more the developing countries which during the past few years have lost some $ 500 billion through trade barriers of various kinds. This must change if ever the developing countries will be able to develop their own scientific and technological capabilities as such NIC-countries as South Korea, Taiwan, Mexico and Brazil have been able to do.

An additional and recent trend which becomes more and more important is the decreasing significance of military science and technology for the strengthening of the national economy. Japan and former West Germany are extraordinary examples of how national economies can be expanded by production for the civilian market alone.

Finally it should be mentioned that the general trend of deregulation and privatization of the markets shifts the power from monopoly to consumer power. As has been mentioned earlier the scientific and technological development must learn to adapt more closely to the needs of the civilian market and financing sector.

In this regard the quest for environmentally sound technologies puts a new pressure on scientific priorities. The restructuring of the industrial economy towards a sustainable development is illustrated by the table below presented by Hazel Henderson in 1990.

There is no question that the local and global environmental degradation will have a strong impact on the future priorities for scientific and technological development. Let us only mention the threats of a global warming due to the emission of various "green-house" gases, especially carbon-dioxide from the burning of fossil fuels and the cutting down of forests.

Restructuring Industrial Economies

Obsolescent Sectors (Unsustainable, Entropic)	Emerging Sectors (Sustainable, low entropy)
* Industries, companies based on heavy use of non renewable energy and materials	* Industries, companies based on efficient use of energy and materials and human skills
* Bureaucratic, large, less flexible	* Entrepreneurial, small, flexible
* Non-recyclable products, packaging manufacturing	* Recyclable products, re-
* Military contracting	* Conservation, innovation
* Products involving toxic, non transit	* Fuel efficient motors, cars, mass
biodegradables polluting materials, throwaway items services	* Solar, renewable energy systems * Communications, information,
* Planned obsolescence	* Infrastructure, education, training
* Chemical pesticides, inorganic fertilizers	* Space communications satellites

* Heavy farm equipment
treaties
* Polluting, inefficient capital equipment,
processes
  process machinery, processing systems
forestation
* Extractive industries with low value
management
  added
prevention
* Fossil fuels, nuclear power generation
* High tech hospital based medical care
* Highly processed foods
prevention
* Advertising encouraging waste and
  polluting practices
* Shopping centre developers

* Peace keeping, surveillance of
* Efficient capital equipment,
* Restorative industries, re-
  desert greening, water quality
* Health promotion and disease
* Organic agriculture, low till systems
* Integrated pest management
* Pollution control, clean up and
prevention
* Natural foods
* Waste recycling and reuse
* Community design and planning
* "Caring" sector

Creativity and the Innovations Process

Creativity in science and technology is of the same nature as creativity in arts although with a somewhat different motivational background. When the pressure on an organization increases to become more and more innovative and responsive to change the organizational environment for human creativity must become more and more flexible.

Individual responsibility and freedom must be allowed to increase if the flow of information into the organization will be transformed in creative use. This in turn will require the change from hierarchical organizations (military) towards self-organizing network organizations. Creativity flourishes best in freedom.

The Innovation process can be divided in three main stages:

the Idea stage;
the Incubation stage;
the Growth stage.

The Innovation Institute in Sweden has divided the Innovation process in some more pragmatic stages for guided product development:

Needs specification;
Idea generation;
Development of innovation ideas;
Pilot production and testing;
Trial production;
Market introduction.

When it comes to the management of the various phrases of the innovation process it might be useful to describe the process by Figure 5 below. In most instances the management style will vary through the different phases as follows:

Phase I   Innovative
                   Idea generation
                   Reflective
                   Discussing
                   Knowledge accumulation
       Management: Creative innovator

Phase II   Activity
                   Risk taking
                   Ability to act
                   Sensitivity to results
                   "Blood, sweat and tears"
       Management: Business entrepreneur

Phase III   Administrative
                   Co-ordinating
                   Law and Order
                   Straight lines
                   Volume, cash-flow
       Management: Administrator

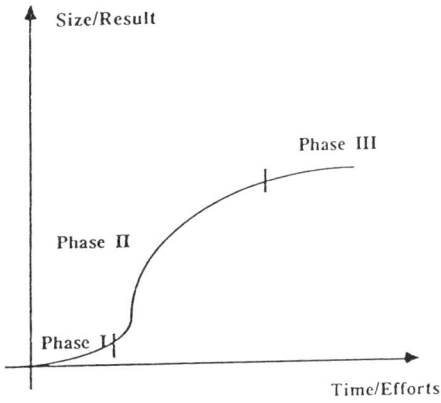

Figure 5

Finally we want to give the following advice to researchers in universities and similar institutions who want to start their own Company:

1. Define clearly the business idea;
2. Have a competent board with market and management capabilities;
3. Have a competent accountant;
4. Have a sustainable risk capital;
5. Work actively with market problems, co-operate with clients;
6. Follow patent development;
7. Have tight economic control, spec. cash-flow;
8. Minimize fixed costs;
9. Always have written agreements;
10. Avoid private security.

The Role of Technology Assessment and Policy Analysis in R&D Decision Making

Bruno Schmitz, Commission of the European Communities, Brussels, Belgium

Introduction

This paper summarizes the lectures given by Bruno Schmitz in two of the sessions of the International Course on Research and Innovation Management (Moscow, 29 October – 8 November 1992). The first session (Thursday, 5 Nov.) was on "The environment of Research, Development and Innovation", the second (Saturday, 7 Nov.) addressed the question of "Financing of Research, Innovation and New Product Development". This summary also incorporates the contribution of B. Schmitz in the Round Table discussion that took place towards the end of the Course.

His interventions, as reflected by the structure of the present summary, concentrated (i) on the "process" of priority-setting in science and technology (S&T), and (ii) on the prospects – and limitations – of technology mapping and related techniques for the identification of research and technology priorities.

These subjects were addressed with particular reference to the European Community (EC) science and technology; that is to say with reference to public intervention in research and technology, and to a quite specific socio-economic and institutional "environment". The methods applied for priority-setting or the prospects of a particular method could be totally different in another context; for example, if the objective was to orient the strategy of a science-based private firm.

It should be noted that these lectures provided a particular perspective on the activities in question based on my own experience as Head of the Strategic Analysis of Science and Technology (SAST) unit of the Directorate General for Science, Research and Development of the Commission of the European Communities (CEC); they should not be taken as indicating the considered opinion of officials within the Commission.

Finally, this paper is only an outline of the presentations that can be found in the documents referred to in footnotes, and in particular in the SAST reports distributed to the Course participants.

The Process of Priority-Setting – A General Model?

The schematic representation of foresight for priority-setting shown in figure 1 is taken from Martin and Irvine, 1989.[1] This "model" relates to priority-determination and does not attempt to describe all types of foresight activities, which could have different functions (direction-setting or anticipatory intelligence, for example). Also the figure does not adequately convey that one or more cycles of foresight may be carried out at different levels/stages (iterations) prior to determining research priorities.

In any case, it should not be viewed as a normative model, nor should it be seen as implying a belief in the possibility of a wholly rational approach to policy-making, but rather as a heuristic model enabling a better understanding to be developed of the nature and role of the different tasks, how the various elements are interrelated and the likely effect on the overall process of a failure to execute successfully any of the main steps involved.

Clearly Strategic Analysis is only one element in the whole foresight process. Neither is it the activity by which overall policy orientation or guidelines can be set, nor should it be considered that a report of the Strategic Analysis should necessarily be the key determinant in the decision to launch or reorient a research programme.

Strategic Analysis at EC level

The mandate of the EC-SAST Unit is consistent with the above; it reads: "to undertake targeted analysis within policy orientations already on, in order to provide a decision base in specific areas of S&T policy".

The provision of a decision base (even if it is not the decision itself) requires one, as shown by the figure, to properly design the process. It must ensure that all major factors, either technical or non-technical, associated with a particular set of S&T issues are adequately considered, and that procedures are in place for consulting and/or informing the various parties concerned on a on-going basis, from the conception of a project, to the diffusion of its results throughout the execution phases. The process must necessarily involve both the performers and funders, as well as the users of the results and other stakeholders (e.g. environmental protection agencies). The first step must be, on the part of the organizations and individuals concerned – policy makers and programme managers – an explicit decision to initiate the process.

Interaction with interested parties

Within the context of the SAST activity, the interaction with the "key-players" is organized as shown in figure 2. While it is relatively easy to set up an appropriate consortium of experts to carry out the desk work and expert consultations and organize the consultation within the institutions (which is the function of the internal steering group), it is much more difficult, in dealing with twelve countries, to ensure proper consultation of the various users and other stakeholders. There are "equivalents", at Community-level, of the national professional associations and interest groups, and also a number of advising Committees of member states representatives which can help on this matter. However, a variety of ad-hoc mechanisms of external consultation are to be developed to complement the existing ones. Networks of experts or industrial representatives may be established to complement the work of the contractors. Each project also normally involves a series of review workshops, in which representatives of relevant professional, industrial and social interests are invited to comment on the analysis which has been conducted and any associated proposals for Community action.

The desk work has also some specific requirements. The need here is to ensure that the selection of contractors is both effective – in finding the appropriate skills for the job – and equitable – in making a fair choice amongst possible contractors who may be located in any member states and have a variety of cultural backgrounds, language skills and specialized expertise. It is to take account of these different cultural backgrounds that any contractor ultimately chosen is asked to not only undertake desk research but also consult as widely as possible with experts in other parts of the Community (or in certain cases outside the Community).

Besides, these more administrative requirements of strategic analysis, there are methodological requirements including:

the need to retrieve and to integrate up-to-date information from a wide variety of sources, whose availability may be limited by commercial or political sensitivity, and which may be expressed in a variety of different forms;

the need for a spectrum of analytical techniques which reflect the diversity of problems to be examined, and which allow reliable forecasts to be made of the development of a scientific or technological area, and the appropriate evaluations to be conducted;

the need to take account of the various interests which may be affected by the issue being studied, and any eventual Community policy response, in a balanced manner.

The latter is particularly challenging and crucial in consideration of the need to reach some kind of "broad consensus" (strict consensus is never achievable) on the most promising S&T options.

Above all, strategic analysis in science and technology must function administratively and methodologically as an integral part of a wider system of policy making tools and procedures. Society's concerns for science and technology cannot be restricted to the practice of research, with the expectation that results will automatically be taken up and used productively in the public interest. The dynamics of science and technology reflect a complex system of interactions in which a range of public policies play a part. Strategic analysis is a tool to help anticipate and monitor this system of interaction, to detect weakness and opportunities for improvement, and to propose appropriate policy responses.

Clearly, however, SAST is just one of a number of "feedback and control" mechanisms available to the Commission and the guidance for RTD policy involves many elements quite outside SAST's competence. In the context of the MONITOR programme itself, there are complementary activities dealing with (i) long-term forecasting (FAST) and (ii) expert evaluation of programmes (SPEAR).

Finally, it is important to stress that the "science" of Strategic Analysis, or more generally the science of "steering science and technology" is relatively immature. The administrative structure for this kind of work must be evolutionary and open, allowing "learning by doing".

Technology-mapping and Related Techniques

In 1991, SAST commissioned a review of technology-mapping and related techniques, focusing on the identification of their potential to assist in the prioritization as related to EC research and technological development[2].

Figure 3 provides a rough summary of the review. Technology-mapping can be seen to comprise a range of techniques, each suitable for providing information or analysis at different stages in technological development, at different taxonomic levels of application of technology and for both firm and national performance.

A priori technology-mapping could help illustrate and analyze the relationships between science, technologies and industrial product groups and the context in which they develop. The techniques reviewed were analyzed in a framework which clarified the analytical distinctions between science and technology and industrial product groups. Mapping the knowledge base showed developments in science and technology and the emergence of networks linking individuals and organizations. Likewise, mapping of artefacts (the physical entities) examined product groups, with consideration of a taxonomy showing relationships between products and broader technological areas. It is useful therefore to bring these techniques together in a user oriented way, providing knowledge of them and their capabilities if not of their detailed operation and mode of application. Also, the use of technology-mapping and related techniques should be seen in conjunction with other techniques providing economic and social perspectives, not in isolation.

Technology-mapping is thus seen to be potentially useful in the system of priority-setting described earlier in this paper, however, this utility must be qualified. RTD priority-setting can be seen at two levels: the selection of technological areas and the determination of options within those areas. The limitations of technology-mapping, particularly its data dependency and subsequent cost effectively rule out its use in the first level of selection. It is felt, however, to be of great value at the second lower level, identifying a number of salient features of a preselected area, including: the distance between firms' and countries' technological capabilities as well as the state of the art and its limits; possible bottlenecks; synergy and complementarity and the nature, dynamic and location of the underlying knowledge base. That is to say that technology-mapping could greatly assist in the setting of priorities within the scope of a predetermined technological agenda.

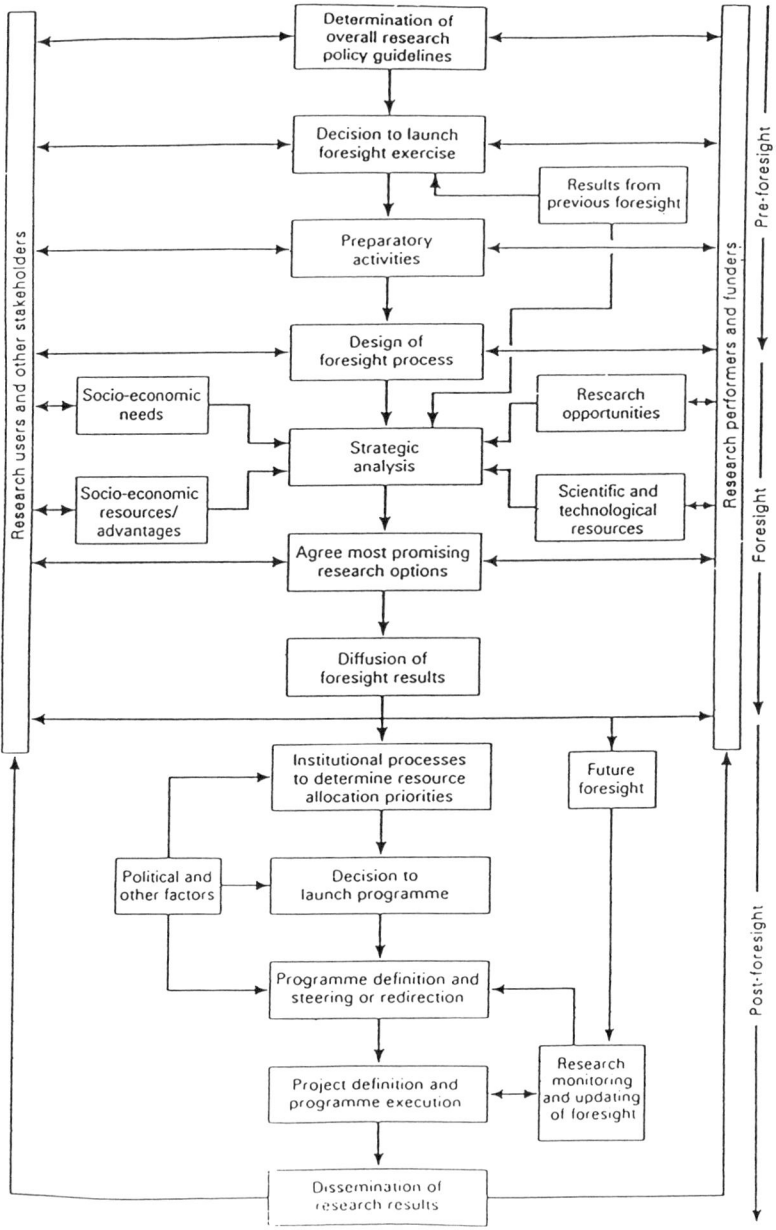

Figure 1. Elements and stages in foresight oriented to priority setting (including implementation)

INTERPLAY BETWEEN PARTICIPANTS IN SAST PROJECTS

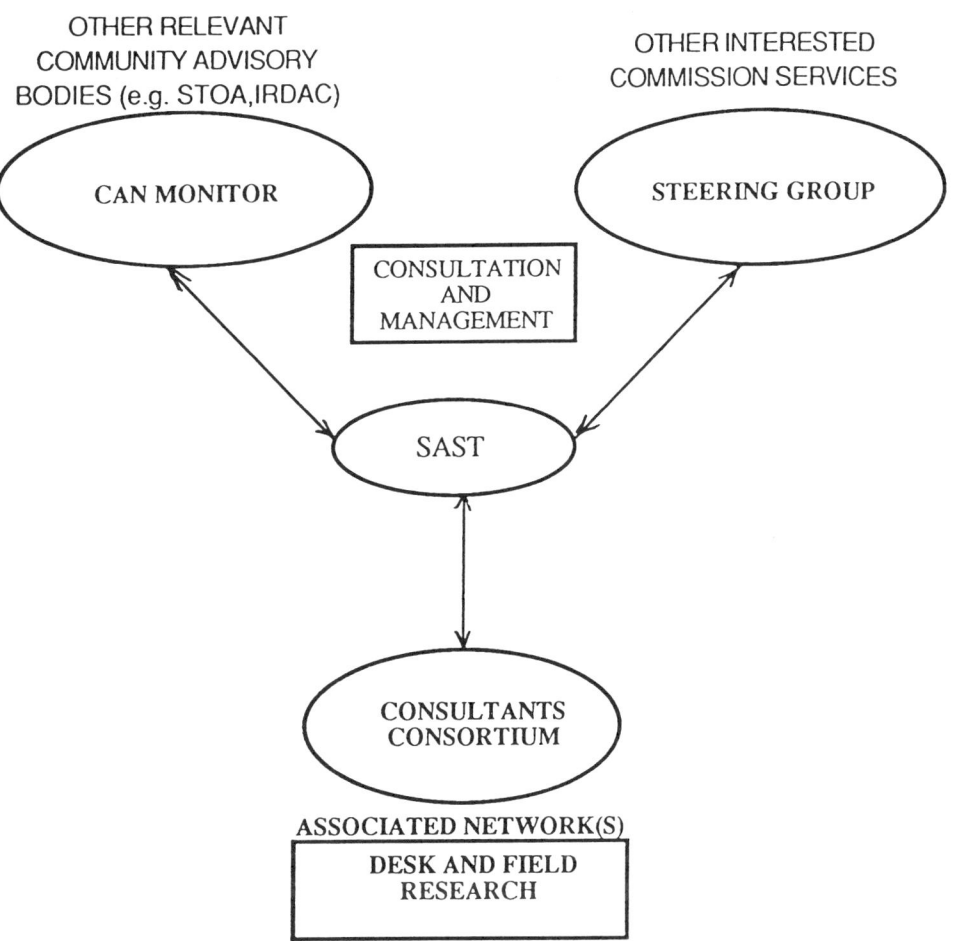

Figure 2

USE / TECHNIQUE	SCIENCE / BASIC RESEARCH STATE	IMPACT	APPLIED RESEARCH/ DEVELOPMENT STATE	IMPACT	TECHNOLOGY ON MARKET STATE	IMPACT	SECTORAL LEVEL	GENERIC TECHNOLOGY	SPECIFIC PRODUCT	FIRM/GROUP PERFORMANCE	NATIONAL PERFORMANCE
TECHNICAL / PERFORMANCE CHARS.			✗		✓		✗	✗	✓	✓	✗
TECHNICAL VS. PERFORMANCE			✗	✗	✓	✓	✓	✓	✓	✗	✗
HEDONIC - TYPE					✓	✓	✗	✗	✓		
QUALITATIVE ARTEFACT MAPPING			✗	✗	✓	✓	✓	✓	✓	✓	✓
BIBLIOMETRIC COUNT	✓		✓				✗	✓		✓	✓
CITATION ANALYSIS	✓	✓	✓	✓			✗	✓		✓	✓
CO-CITATION ANALYSIS	✓	✓	✓	✓			✗	✓		✓	✓
CO-WORD ANALYSIS	✓	✓	✓	✓			✓	✓		✓	✓
CO-NOMINATION ANALYSIS	✓	✓	✓	✓			✓	✓		✓	✓
PATENT COUNT	✓	✓	✓	✓	✗	✗	✓	✓	✓	✓	✓
PATENT CONTENT ANALYSIS	✓	✓	✓	✓	✗	✗	✓	✓	✓	✓	✓
QUALITATIVE KNOWLEDGE MAPPING	✓	✓	✓	✓	✗	✗	✓	✓	✓	✓	✓
COMBINATION TECHNOLOGY AND ARTEFACT	✓	✓	✓	✓	✓	✓	✓	✓	✓	✓	✓

√ – Indicates where technique may be of particular value
x – Indicates where technique is of limited value

Figure 3

References

1 Martin, B.R. and Irvine, J. (1989), "Research Foresight: Priority-Setting in Science", London, Pinter Publishers.

2 Mark Boden, CREST, "The Identification of Technology Priorities for European Research and Technology Development: A Review of Technology-mapping and related techniques", SAST Project No 5, February 1992 (CD-NA-14722-EN-C).

Managing the Innovative R&D Team

Giovanni Abramo, Italian National Research Council (CNR), Department of Innovation Transfer, Rome, Italy

Introduction to the Articles

In the context of the "Human and Organizational Resources" session of the course, the lecture has focused on the technology-based innovation process in terms of critical behavioural functions.

The readings proposed to the participants included an article by Edward B. Roberts and Alan R. Fusfeld and a case study.

The former is a report of findings of over fifteen years of studies conducted mainly at the Massachusetts Institute of Technology in the field of innovation and R&D management.

The article greatly contributes to filling a gap in the theory of organization for innovation. Examination of the innovation process in numerous industrial companies in the U. S. and Europe indicates that many attempts to innovate suffer from ineffective management and inadequately staffed organizations. The authors' and other scholars' research has accumulated precious knowledge about the activities that are requisite to innovation and the characteristics of the people who perform these activities most effectively. The awareness of such findings should be of utmost importance to R&D leaders and policy makers in the scientific and technological fields.

While 70 to 80 percent of technical effort in organizations falls into a routine problem-solving category, in their article the authors emphasize the 20 to 30 percent which is unique and critical in the innovation process.

The authors identify and describe five informal but critical behavioural functions, which are needed for effective execution of technology-based innovative projects. They observe that some individuals are capable of performing concurrently more than one of these critical functions. At the same time, some roles frequently need to be fulfilled by more than one person. They also find that person's role in the innovation process may change over the course of his or her career.

While the five roles arise in differing degrees in each of the several steps involved in the technology-based innovation process, the inadequate performance of any of the critical functions will certainly reduce the innovative effectiveness, sometimes to the point of affecting the project's success.

The authors also discuss the managerial implications of their findings, particularly with respect to manpower planning, job design, objective setting, and performance measurement and rewards. They then illustrate how an organizational assessment can be carried out. Finally, they discuss how critical functions concepts may be appropriate in other types of organizations.

The case study deals with a research-based innovation project. It is a perfect complement to the theory presented in the article. In fact, the situation portrayed in the case elicits deep thinking on the behavioural roles fulfilled by the team members, the formal and informal communication links, and the management approach, both at lab and project levels.

The Sojourner at Simmons

by Dr. Alex Bavelas, modified by Harold J. Leavitt

Bob Knowlton was sitting along in the conference room of the laboratory. The rest of the group had gone. One of the secretaries had stopped and talked for a while about her husband's coming induction in the Army, and finally left. Bob slid a little further down in his chair, looking with satisfaction at the results of the first test run of the photon unit.

He liked to stay after the others had gone. His appointment as project head was still new enough to give him a deep sense of pleasure. His eyes were on the graphs before him, but in his mind he could hear Dr. Jerrold, the project director, saying again, "There's one thing about this place that you can bank on. The sky is the limit for a man who can produce!" Knowlton felt again the tingle of happiness and embarrassment. Well, dammit, he said to himself, he had produced. He wasn't kidding anybody. He had come to the Simmons Laboratories two years ago. During routine testing of some rejected Clanson components he had stumbled on the idea of the photon correlator, and the rest had just happened. Jerrold had been enthusiastic; a separate project had been set up for further research and development of the device, and he had gotten the job of running it. The whole sequence of events still seemed a little miraculous to Knowlton.

He shrugged out of the reverie and bent determinedly over the sheets when he heard someone come into the room behind him. He looked up expectantly. Jerrold often stayed late himself, and now and then dropped in for a chat. This always made the day's end especially pleasant for Bob. It was Jerrold, accompanied by a stranger. The stranger was tall, thin, and rather dark. He wore steel-rimmed glasses and had on a very wide leather belt with a large brass buckle.

Jerrold introduced him: "Bob, this is Simon Fester. I'm glad we found you in. I've got to rush, but if you've got a few minutes, I wonder if you could chat with Simon. We were talking about your work and he's very much interested in it. I thought it would be better if he got a first-hand idea of what you are doing." Jerrold excused himself. Bob waved Fester to a chair. Fester didn't seem to belong in any of the standards categories of visitors, clients, visiting firemen, or stockholders. Bob pointed the sheets on the table. "These are the preliminary results of a test we're running. We've got a new gadget by the tail and we're trying to understand it. It's not finished, but I can show you the section that we're testing." He stood up, but Fester was deep in the graphs. After a moment he looked up with an odd grin. "These look like plots of a Jennings surface. I've been playing around with some auto-correlation functions of surfaces – you know the stuff." Bob, who had no idea what he was referring to, grinned back, nodded, and immediately felt uncomfortable. "Let me show you the monster," he said, and led the way to the work room.

After Fester left, Knowlton slowly put the graphs away, feeling vaguely annoyed. Then, as if he had made a decision, he quickly locked up and took the long way out so that he would pass Jerrold's office. But the office was locked. Knowlton wondered whether Jerrold and Fester had left together.

The next morning Knowlton dropped into Jerrold's office, mentioned that he had talked with Fester, and asked who he was.

"Sit down for a minute," Jerrold said, "I want to talk to you about him. What do you think of him?" Knowlton replied truthfully that he thought. Fester was very bright and probably very competent. Jerrold looked pleased.

"We're taking him on," he said. "He's had a very good background in a number of laboratories, and he seems to have ideas about the problems we're tackling here." Knowlton nodded in agreement, instantly wishing that Fester would not be placed with him.

"I don't know yet where he will finally land," Jerrold continued, "but he seems interested in what you are doing. I thought he might spend a little time with you by way of getting started," Knowlton nodded thoughtfully. "If his interest in your work continues, you can add him to your group."

"Well, he seemed to have some good ideas even without knowing **exactly** what we are doing," Knowlton answered. "I hope he stays; we'd be glad to have him."

Knowlton walked back to the lab with mixed feelings. He told himself that Fester would be good for the group. He was no dunce, he would produce. Knowlton thought again of Jerrold's promise when he had promoted him. – "the man who produces gets ahead in this outfit." The words seemed to him to carry overtones of a threat now.

The next day, Fester didn't appear until mid-afternoon. He explained that he had had a long lunch with Jerrold; discussing his place in the lab. "Yes," said Knowlton, "I talked with Jerry this morning about it, and we both thought you might work with us for a while."

Fester smiled in the same knowing way that he had smiled when he mentioned the Jennings surfaces. "I'd like to", he said.

Knowlton introduced Fester to the other members of the lab. Fester and Link, the mathematician of the group, hit it off well together, and spent the rest of the afternoon discussing a method of analysis of patterns that Link had been working on over the month.

It was six-thirty when Knowlton finally left the lab that night. He had waited almost eagerly for the end of the day to come – when they would all be gone and he could sit in the quiet room, relax, and think it over. "Think what over?" he asked himself. He didn't know. Shortly after five they had all gone except Fester and what followed was almost a duel. Knowlton was annoyed that he was being cheated out of his quiet period and finally resentfully determined that Fester should leave first.

Fester was sitting at the conference table reading and Knowlton was sitting at his desk in the little glass-enclosed cubby that he used during the day when he needed to be undisturbed. Fester had gotten the last year's progress reports out and was studying them carefully. The time dragged. Knowlton doodled on a pad, the tension growing inside of him. What did Fester think he was going to find in the reports?

Knowlton finally gave up and they left the lab together. Fester took several of the reports with him to study in the evening. Knowlton asked him if he thought the reports gave a clear picture of the lab's activities.

"They're excellent," Fester answered with obvious sincerity. "They're not only good reports; what they report is damn good, too!" Knowlton was surprised at the relief he felt and grew almost jovial as he said goodnight.

Driving home, Knowlton felt more optimistic about Fester's presence in the lab. He had never fully understood the analysis that Link was attempting. If there was anything wrong with Link's approach, Fester would probably spot it. "And, if I'm any judge," he murmured, "he won't be especially diplomatic about it."

He described Fester to his wife, who was amused by the broad leather belt and the brass buckle.

"It's the kind of belt the Pilgrims must have worn," she laughed.

"I'm not worried about how he holds his pants up," he laughed with her. "I'm afraid that he's the kind that just have to make like a genius twice each day. And that can be pretty rough on the group."

Knowlton had been asleep for several hours when he was jerked awake by the telephone. He realized it had rung several times. He swung off the bed muttering about damn fools and telephones. It was Fester. Without any excuses, apparently oblivious of the time, he plunged into an excited recital of how Link's patterning problem could be solved.

Knowlton covered the mouthpiece to answer his wife's stage whisper, "Who is it?"

"It's the genius."

Fester, completely ignoring the fact that it was two in the morning, proceeded in a very excited way to start in the middle of an explanation of a completely new approach to certain of the photon lab problems that he had stumbled on while making analysis of some past experiments. Knowlton managed to put some enthusiasm in his own voice and stood there, still half-dazed and very uncomfortable, listening to Fester talk endlessly, it seemed to him, about what he had discovered, which was only probably a new approach, but also an analysis which showed how inherently weak the previous experiment had been and how experimentation along that line would certainly have been inconclusive.

The following morning he spent the entire morning with Fester and Link, the mathematician, the weekly group meeting had been called off so that Fester's work of the previous night could be gone over intensively. Fester was very anxious that this be done and Knowlton was not too unhappy to call the meeting off for reasons of his own.

For the remainder of the day, Fester sat in the back office that had been turned over to him and did nothing but scan the progress reports of the work that had been done in the last six months. Knowlton caught himself feeling apprehensive about the reaction that Fester might have to some of this work.

He was a little surprised at his own feelings. He had always been proud – although he had put on a convincingly modest face of the rate at which new ground in the study of photon-measuring devices had been broken in his group. Now he wasn't sure, and it seemed to him that Fester might easily show that the line of research they had been following was unsound or even unimaginative.

The next morning, as was the custom in Bob's group, the members of the lab, including the girls, sat around the table in the conference room for a group meeting. Bob always prided himself on the fact that the work of the lab was guided and evaluated by the group as a whole. He was fond of repeating that it is not a waste of time to include secretaries in such meetings because often what started out as a boring explanations of fundamental assumptions to a naive listener uncovered new ways of regarding these assumptions that would not have occurred to the lab member who had long ago accepted them as a necessary basis for the research he was doing.

These group meetings also served Bob in another sense. He admitted to himself that he would have felt far less secure if he had had to direct the work out of his own mind, so to speak. With the group meeting as the principle of leadership it was always possible to justify the exploration of blind alleys as valuable because of the general educative effect of the team. Fester, Lucy, and Martha were there; Link was sitting next to Fester, their conversation concerning Link's mathematical study apparently continuing from yesterday. The other members – Bob Davenport, George Thurlow, and Arthur Oliver – were there and waiting quietly.

Knowlton, for reasons that he didn't quite understand, proposed for discussion this morning a problem that they had previously spent a great deal of time discussing and had come to the conclusion that a solution was impossible, that there was no feasible way of treating it in an experimental fashion. When Knowlton proposed the problem, Davenport remarked that there was hardly any use going over it again; he was satisfied that there was no way of approaching the problem with the equipment and the physical capacities of the lab.

This statement had the effect of a shot of adrenalin on Fester. He said he would like to know what the problem was in detail, and walking to the blackboard, began setting down "the factors" as various members of the group began both discussing the problem and simultaneously listing the reasons why it had been abandoned. Very early in the description of the problem it was evident that Fester was going to disagree about the impossibility of attacking it. The group realized this and finally the descriptive materials and their recounting of the reasoning that had led to its abandonment dwindled away. Fester began his statement, which, as it proceeded, might well have been prepared the previous night although Knowlton knew this was impossible. He couldn't help being impressed with the organized and logical way that Fester was presenting ideas that must have occurred to him only a few minutes before.

Fester had some things to say, however, which left Knowlton with a mixture of annoyance, irritation, and, at the same time, a rather smug feeling of superiority over Fester in at least one area. Fester was of the opinion that the way the problem had been analyzed was really typical of what happened when such thinking was attempted by a group. With an air of sophistication which made it difficult for a listener to dissent, he proceeded to make general comments on the American emphasis on team ideas, satirically describing the ways in which they led to a "high level of mediocrity."

During this time Knowlton observed that Link stared studiously at the floor and he was very conscious of George Thurlow's and Bob Davenport's glances toward him at several points of Fester's little speech. Inwardly, Knowlton couldn't help feeling that this was one point at least in which Fester was off on the wrong foot. The whole lab, following Jerry's lead, talked, if not practised, the theory of small research teams as the basic organization for effective research. Fester insisted that the problem could be approached and he would like to study it for a while himself.

Knowlton ended the morning session by remarking that the meetings would continue and the very fact a supposedly insoluble experimental problem was now going to get another chance was indication of the value of such meetings. Fester immediately remarked that he was not at all averse to meetings for the purpose of informing the group of the progress of its members that the point he wanted to make was that creative advances were seldom accomplished in such meetings, that they were made by the individual "living with" the problem closely and continuously, a sort of personal relationship to it. Knowlton went to say to Fester that he was very glad Fester had raised these points and he was sure the group would profit by re-examining the basis on which they had been operating. Knowlton agreed that individual effort was probably the basis for making major advances, but that he considered the group meetings useful primarily because of the effect they had on keeping the group together and on helping the weaker members of the group keep up with the advances of the ones who were able to progress more easily and quickly along the analysis of problems.

It was clear, as the weeks went by and meetings continued as they did, that Fester came to enjoy them because of the pattern which the meetings assumed. It became almost typical for Fester to hold forth and it was also clear that he was, without questions, more brilliant, better prepared on the various subjects which were related to the problems being studied, more capable of going ahead than anyone there. Knowlton grew increasingly disturbed as he realized that his leadership of the group had been, in fact, taken over. In Knowlton's occasional meetings with Dr. Jerrold, whenever the subject of Fester was mentioned, he would comment only on the ability and obvious capacity for work that Fester had, somehow never feeling that he could mention his own part, but also because it was quite clear that Jerrold himself was considerably impressed with Fester's work and with the contacts he had outside the photon laboratory.

Knowlton at this time began to feel that the intellectual advantages that Fester had brought to the group perhaps did not quite compensate for what he felt were evidences of breakdown of the co-operative group spirit which he had seen before Fester's coming. More and more of the morning meetings were skipped. Fester's opinion concerning the abilities of others of the group, with the exception of Link, was obviously low and at times during morning meetings, or in smaller discussions, he had been on the point of rudeness, refusing at certain times to pursue an argument when he claimed it was based on ignorance of the other person of the facts involved. His impatience of the others led him to make remarks of this kind also to Dr. Jerrold. This Knowlton inferred from a conversation he had had with Jerrold in which Jerrold had been asking whether Davenport and Oliver were going to be continued on. The fact that he did not mention Link, the mathematician, led Knowlton to feel that this was the result of conversations that Fester had had privately with Jerrold.

It was not difficult for Knowlton to make a quite convincing case on the question of whether for brilliance of Fester was actually a sufficient recompense for the beginning of this breaking up of the group. He took the opportunity to speak privately with Davenport and Oliver and it was quite clear that both of them were uncomfortable with the relationship with Fester. Knowlton didn't press the discussion beyond the point of hearing them say in one way or another, that they did sometimes felt awkward and it was sometimes difficult for them to understand the arguments he advanced, but often embarrassing to ask him to fill in the background on which he felt such arguments were valid. Knowlton did not interview Link in this manner.

About six months after Fester's coming into the photon lab, a meeting was scheduled in which the sponsors of much of the research going on were coming to get some ideas of the work and its progress. It was customary, at these meetings, for project heads to present the research being conducted in their groups. The members of each laboratory group were invited to other meetings which were held later in the day and open to all, but the special meetings were usually made up only of project heads, the head of the laboratory, and the sponsors.

As the time for the special meeting approached, it seemed to Knowlton that he must avoid the presentation at all cost. His reasons for this were that he could not trust himself to present the ideas that Fester had advanced and on which some work had been done, because of his apprehension as to whether he could present them in sufficient detail and answer such questions about them as might be asked. On the other hand, he did not feel he could ignore these newer lines of work and present only the material which had been done or had been started before Fester's arrival, which he was perfectly competent to do. He felt also that it would not be beyond Fester, in his blunt and undiplomatic way, if he were present at the meeting, to make comment on his own presentation and reveal the inadequacy which Knowlton felt he had. It seemed quite clear it would not be easy to keep Fester from attending the meeting in spite of the fact that he was not on the administrative level which was invited.

Knowlton found an opportunity to speak to Jerrold and raised the question. He remarked to Jerrold that, of course, with the meetings coming up, with the interest in the work, and with the contributions that Fester had been making, he would probably like to come to these meetings, but there was a question of the feelings of the others in the group if Fester alone were invited. Jerrold passed this over lightly by saying that he didn't think the group would fail to understand Fester's rather difficult position and he thought that Fester, by all means, should be invited. Knowlton then immediately said that he had thought so too and that he felt Fester should present the work because much of it was what he had done; and that this would be a nice way to recognize Fester's contributions and to reward him since he was eager to be recognized as a productive member of the lab. Jerrold agreed and so the matter was decided.

Fester's presentation was very successful and in some ways dominated the meeting. He attracted the interest and attention of many of those who had come and following his presentation the questions persisted for a long period. Later in the evening at the banquet, at which the entire laboratory force was present, in the cocktail period before the dinner, a little circle of people had formed about Fester, one of them being Jerrold himself, and the discussion concerning the application of the theory he was proposing went in with great interest. All of this disturbed Knowlton and his reaction and behaviour were characteristic. He joined the circle, praised Fester to Jerrold and others, and remarked how able and how brilliant some of his work was.

Knowlton, without consulting anyone, began to take some interest in the possibility of a job elsewhere. After a few weeks he found that a new laboratory of considerable size was being organized in a nearby city, and that the kind of training he had would enable him to get a project head job equivalent to the one he had at the lab, with slightly more money.

He immediately accepted it and notified Jerrold, by letter, which he mailed on a Friday night to Jerrold's home. The letter said that he had found a better position; that there were personal reasons why he didn't want to appear at the lab anymore; that he would be glad to come back at some future time from his new job, some four miles away, to assist if there was any mix-up in the work. It stated that he felt sure Fester could supply any leadership that was required for the group, and that his decision to leave so suddenly was based on personal problems. It hinted at problems of health of his family, his mother, and father. All of this was fiction, of course. Jerrold took it at face value, but still felt that this was very strange behaviour and quite unaccountable since he had always felt his relationship with Knowlton had been warm and that Knowlton was satisfied, quite happy, and productive.

Jerrold was considerably disturbed because he had already decided to place Fester in charge of another project that was going to be set up very soon and had been wondering how to explain this to Knowlton in view of the obvious help, assistance, and value Knowlton was getting from Fester and the high regard in which he held him. As a matter of fact, he had considered the possibility that Knowlton could add to his staff another person with the kind of background and training that had been unique in Fester and had proved so valuable.

Jerrold did not make any attempt to meet Knowlton. In a way he felt aggrieved about the whole thing. Fester, too, was surprised at the suddenness of Knowlton's departure and when Jerrold, in discussing this, asked him whether he preferred to stay with the photon group instead of the project for the Air Force which was being organized, he chose the Air Force Project and went on to that job the following week. The photon lab was hard hit. The leadership of the lab was given to Link with the understandings that this would be temporary until someone could come in to take over.

Staffing the Innovative Technology-Based Organization

by Edward B. Roberts and Alan R. Fusfeld

In this article, the authors identify and describe five informal but critical behavioural functions, which are needed for effective execution of technology-based innovative projects. They observe that some individuals are capable of performing concurrently more than one of these critical functions. They also find that a person's role in the innovation process may change over the course of his or her career. The authors discuss the managerial implications of their findings, particularly with respect to manpower planning, objective setting, and performance measurement and rewards. They then illustrate how an organizational assessment can be carried out. Finally, they discuss how critical functions concepts may be appropriate in other types of organizations.

This article examines the technology-based innovation process in terms of certain behavioural functions. These functions are usually informal, but they are critical. They can be the key to an effective organizational base for innovation. This approach to the innovation process is similar to that taken by early industrial theorists, such as Frederick W. Taylor who focused on organizing efficiently the production process. However, examination of how industry has organized its innovation tasks – those tasks needed for product or process development and for responses to non routine demands – indicates an absence of comparable theory. Many corporations' attempts to innovate consequently suffer from ineffective management and inadequately staffed organizations. Yet, through studies conducted largely in the last fifteen years, we now know much about the activities that are requisite to innovation. We also know about the characteristics of the people who perform these activities most effectively.

The Innovation Process

The major steps involved in the technology-based innovation process are shown in Figure 1. Although the project activities do not necessarily follow cash other in a linear fashion, there is more or less clear demarcation between them. Each stage and its activities, moreover, require a different mix of "people" skills and behaviours to be carried out effectively.

The figure portrays six stages as occurring in the typical technical innovation project. It also shows sixteen representative activities that are associated with innovative efforts. The six stages are identified as:

1. Preproject;
2. Project possibilities;

3. Project initiation;
4. Project execution;
5. Project outcome evaluation;
6. Project transfer.

These stages often overlap and frequently recycle.[1] For example, problems of findings that are generated during project execution may cause a return to project initiation activities. Outcome evaluation can restart additional project execution efforts. And, of course, project cancellation can occur during any of these stages, thus redirecting technical endeavours back into the preproject phase.

A variety of different activities are undertaken during each of the six stages. Some of the activities, such as generating new technical ideas, arise in all innovation project stages, from preproject to project transfer. Our research studies and consulting efforts in dozens of companies and government labs, however, have shown other activities to be concentrated in specific stages, as discussed below.

Figure 1. A Multi-stage View of a Technical Innovation Project.

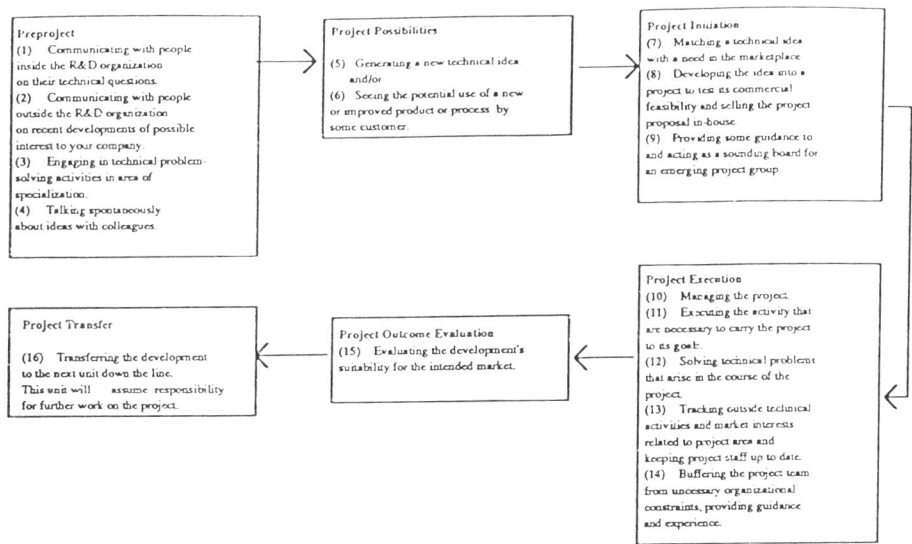

Preproject. Before formal project activities are undertaken in a technical organization, considerable technical work is done, which provides a basis for later innovation efforts. Scientists, engineers, and marketing people find themselves involved in discussions that are internal and external to the organization. Ideas are discussed in rough-cut ways and broad parameters of innovative interest are established. Technical personnel work on problem-solving efforts to advance their own areas of specialization. Our discussions with numerous industrial firms in the U.S. and Europe suggest that from 30 to 60 percent of all technical effort is devoted to work outside of or prior to formal project initiation.

Project possibilities. Specific ideas for possible projects arise from the preproject activities. They may be technical concepts for developments, they may also be perceptions of possible customer interests in product or process changes. Customer-oriented perspectives may originate with technical, marketing, or managerial personnel, who develop these ideas out of their own imaginations or from direct contact with customers or competitors. Recent evidence indicates that many of these ideas enter as "proven" possibilities inasmuch as these ideas have already been developed by the customers themselves.[2]

Project Initiation. As ideas evolve through technical and marketing discussions and exploratory technical efforts, the innovation process moves into a more formal project initiation stage. Activities occurring during this phase include attempts to match the directions of technical work with perceived customer needs. (Of course, such customer needs may exist either in the production organization or in the project marketplace.) Inevitably, a specific project proposal is written, proposed budgets and schedules are produced, and informal pushing as well as formal presentations are undertaken in order to sell the project. A key input during this stage is the counselling encouragement that senior technical professionals or lab and marketing management may provide to the emerging project team.

Project execution. When the project is approved formally, activities increase in intensity and focus. Usually, someone undertakes planning, leadership, and co-ordinating efforts. These efforts are related to the many continuing activities of the engineers and scientists assigned to the project. These activities include problem solving and the generation of technical ideas. Technical people make special attempts to monitor (and transfer in) the results of previous activity as well as relevant external information. Management or marketing people take a closer look at competitors and customers to be sure the project is appropriately targeted.[3] Senior people try to protect the project from being cut off prematurely. The project manager and other enthusiasts fight to defend their project's virtues (and budget). Unless cancelled, the project continues toward completion of its objectives.

Project Outcome Evaluation. When technical effort seems complete, most projects undergo another intense evaluation to see how the results compare with prior expectations and current market perceptions. If successful innovation is to occur, some further implementation must take place. The interim results are either transferred to manufacturing (where they are either embodied in the manufacturing process or produced in volume) or transferred to further stages of development. All such later stages involve heavier expenditures. The project outcome evaluation can then be viewed as a way to screen projects prior to their possible transfer into these later stages.

Project Transfer. If the project results survive this evaluation, transfer efforts take place (e.g., from central research to product department R&D, or from development to manufacturing engineering).[4] The project's details may require further technical documentation to facilitate the transfer. Key technical people may be shifted to the downstream unit to transfer their expertise and enthusiasm, since downstream staff members in the technical or marketing areas often need instruction to assure effective continuity. Within the downstream organizational unit, the cycle of stages may begin again, perhaps by passing the earliest two stages and starting with project initiation or even project execution. This "pass down" continues until successful innovation is achieved, unless project termination occurs first.

Needed Roles

Assessment of activities involved in the several-stage innovation process, as just described, points out that the repeated direct inputs of five different work roles are critical to innovation. The five roles arise in differing degrees in each of several steps. Furthermore, different innovation projects obviously call for variations in the required role mix at each stage. Nevertheless, all five work roles must be carried out by one or more individuals if the innovation is to pass effectively through all six steps. The five critical work functions are:

Idea Generating: Analysing or synthesizing information about markets, technologies, approaches, or procedures, from which is generated an idea for a new technical approach or procedure, or a solution to a challenging technical problem.[5] The analysis or synthesis may be implicit or explicit; the information may be formal or informal.

Entrepreneuring or Championing: Recognizing, proposing, pushing, and demonstrating a new technical idea, approach or a procedure for formal management approval.[6]

Project Leading: Planning and co-ordinating the diverse sets of activities and people involved in moving a demonstrated idea into practice.[7]

Gate keeping: Collecting and channelling information about important changes in the internal and external environments. Information gate keeping can be focused on developments in the market, in manufacturing, or in the world of technology.[8]

Sponsoring or Coaching: Guiding and developing less experienced personnel in their critical roles; behind-the-scenes support, protection, advocacy, and sometimes "bootlegging" of funds.[9]

Lest the reader confuse these roles as mapping one-for-one with different people, three points need emphasis: (1) some roles, e.g., idea generating, frequently need to be fulfilled by more than one person in a project team in order for the project to be successful; (2) some individuals occasionally fulfil more than one of the critical functions; (3) the roles that people play periodically change over a person's career with an organization. The latter two points will be discussed in more depth later in this article.

Critical functions

These five critical functions represent the various roles that must be carried out for successful innovation to occur. They are critical from two points of view. First, each role is unique and demands unique skills. A deficiency in any one of the roles contributes to serious problems in the innovation effort, as we shall illustrate below. Second, each role tends to be carried out primarily by relatively few individuals, thereby making the critical role players even more unique. If any one of these individuals leaves, the problem of recruiting a replacement is very difficult. The specific qualities needed in the replacement usually depend on unstated role requirements. Most critical functions cannot be fulfilled by new recruits to an organization.

We must add at this point that another role clearly exists in all innovative organizations, but it is not an *innovative* role. "Routine" technical problem solving must be carried out in order to advance innovative efforts. Indeed, the vast bulk of technical work is probably routine. It requires professional training and competence, to be sure, but it is nonetheless routine in character for an appropriately prepared individual. A large number of people in innovative organizations do very little critical functions work; others who are important performers of the critical functions also spend a good part of their time in routine problem-solving activity. Our estimate, supported now by data from numerous organizations, is that 70 to 80 percent of technical effort falls into this routine problem-solving category. But, the 20 to 30 percent that is unique and critical is the part we emphasize here.

Generally, the critical functions are not specified within job descriptions, since they tend to fit neither administrative nor technical hierarchies. But they represent necessary activities for R&D, such as problem definition, idea nurturing, information transfer, information integration, and program pushing. Consequently, these role behaviours are the underlying informal functions that an organization carries out as part of the innovation process. Beyond the five roles described earlier, different business environments may also demand that additional roles be performed in order to assure innovation.[10]

It is desirable for every organization to have a balanced set of abilities for carrying out these roles as needed. Unfortunately, few organizations have such a balanced set. Some organizations overemphasize one role (e.g., idea generating) and underplay another role (e.g., entrepreneuring). Technical organizations tend to assume that the necessary set of activities will somehow be performed. As a consequence, R&D labs often lack sensitivity to the existence and importance of these roles, which, for the most part, are not defined within the formal job structure. Yet, the way in which critical functions are encouraged and made a conscious part of technology management is probably an organization's single most important area of leverage for maintaining and improving effective innovation.

Impact of Role Deficiencies

Such an analytic approach to developing an innovative team has been lacking in the past. Consequently, many organizations suffer because one or more of the critical functions are not being performed adequately. Certain characteristic signs can provide evidence that a critical function is missing.

Idea generating is deficient if the organization is not thinking of new and different ways of doing things. However, when a manager complains of insufficient ideas, we commonly find the real deficiency to be that people are not aggressively entrepreneuring or championing ideas – either their own or others. Pools of unexploited ideas that seldom come to managers' attention are evidence of an entrepreneuring deficiency.[11]

Project leading is suspect if schedules are not met, activities "fall through cracks" (e.g., co-ordinating with a supplier), people do not have a sense of the overall goal of their work, or units that are needed to support the work back out of their commitments. Project leading is most commonly recognized by the formal appointment of a project manager. In research, as distinct from development, this formal role is often omitted.

Gate keeping is inadequate if news of changes in the market, technology, or government legislation comes without warning. It is also inadequate if needed information is not passed along to people within the organization. If, six months after the project is completed, you suddenly realize that you have just succeeded in reinventing a competitor's wheel, your organization is deficient in needed gate keeping! Gate keeping is further lacking when the wheel is invented just as a regulatory agency outlaws its use.

Inadequate or inappropriate sponsoring or coaching often explains how projects are pushed into application too soon, why project managers have to spend too much time defending their work. It also explains why personnel complain that they do not know how to "navigate the bureaucracy" of their organizations.

The importance of each critical function varies with the development stage of the project. Initially, idea generation is crucial. Later, entrepreneurial skill and commitment are needed to develop the concept into a viable activity. Once the project is established, good project leadership is needed to guide its progress. Of course, the need for a critical function does not abruptly appear and disappear. Instead, the need grows and diminishes. Each function is the focus at some points, but it is of lesser importance at others. Thus, the absence of a function at a time when it is potentially important is a serious weakness, regardless of whether or not the role had been filled at an earlier, less crucial time. As a corollary, assignment of an individual to a project, at a time when the critical role that he or she provides is not needed, leads to frustration for the individual and to a less effective project team.

Frequently, we have observed that personnel changes that occur because of career development programmes often remove critical functions from a project at a crucial time. Although these roles are usually performed informally, job descriptions are made in terms of technical specialities. Thus, personnel replacements are chosen on the basis of their technical qualifications rather than on their ability to fill the needs of the vacated critical roles. Consequently, the project team's innovative effectiveness is reduced, sometimes to the point of affecting the project's success.

Characteristics of the Role Players

Compilation of several thousand individual profiles of staff in R&D and engineering organizations has demonstrated patterns in the characteristics of the people who performs each innovation function.[12] These patterns are shown in Table 1. The table indicates which persons are predisposed to be more interested in one type of activity than another and to perform certain types of activities well. For example, a person who is comfortable with abstractions and theory might feel more suited to the idea-generating function than would someone who is very practical. In any unit of an organization, people with different characteristics can work to complement each other. For instance, a person who is effective at generating ideas can be teamed with a colleague who is good at gate keeping and with another colleague who has good entrepreneurial abilities. Of course, each person must understand his or her own expected role in a project and must appreciate the roles of others in order for the teaming process to be successful. As will be discussed later, some people have sufficient breadth to perform well in multiple roles.

Table 1 underlies our conclusion that each of the several roles required for effective technical innovation presents unique challenges and must be filled with different types of people. Each type must be recruited, managed, and supported

Table 1. Critical Functions in the Innovation Process.

Critical Function	Personal Characteristics	Organizational Activities
Idea Generating	Expert in one or two fields. Enjoys conceptualization; comfortable with abstractions. Enjoys doing innovative work. Usually is an individual contributor. Often will work alone.	Generates new ideas and tests their feasibility. Good at problem solving. Sees new and different ways of doing things. Searches for the breakthroughs.
Entrepreneuring or Championing	Strong application interests. Possesses a wide range of interests. Less propensity to contribute to the basic knowledge of a field. Energetic and determined; puts self on the line.	Sells new ideas to others in the organization. Gets resources. Aggressive in championing his or her "cause". Takes risks.
Project Leading	Focus for decision making, information and questions. Sensitive to the needs of others. Recognizes how to use the organizational structure to get things done. Interested in a broad range of disciplines and in how they fit together (e.g. marketing, finance).	Provides the team leadership and motivation. Plans and organizes the project. Insures that administrative requirements are met. Provides necessary co-ordination among team members. Sees that the project moves forward effectively. Balances the project goals with organizational needs.
Gate keeping	Possesses a high level of technical competence. Is approachable and personable. Enjoys the face-to-face contact of helping others.	Keeps informed of related developments that occur outside the organization through journals, conferences, colleagues, other companies. Passes information on to others; finds it easy to talk to colleagues. Serves as an information resource for others in the organization (i.e. authority on whom to see or on what has been done). Provides informal co-ordination among personnel.
Sponsoring or Coaching	Possesses experience in developing new ideas. Is a good listener and helper. Can be relatively objective.	Helps develop people's talents. Provides encouragement, guidance, and acts as a sounding board for the project leader and others. Often is a more senior person who provides access to a power base knows the organizational ropes. within the organization – a senior person. Buffers the project team from unnecessary organizational constraints. Helps the project team to get what it needs from the other parts of the organization. Provides legitimacy and organizational confidence in the project.

differently; offered different sets of incentives; and supervised with different types of measures and controls. However, most technical organizations seem not to have grasped this concept. The result is that all technical people tend to be recruited, hired, supervised, monitored, evaluated, and encouraged as if their principal roles were those of creative scientists, or, worse yet, of routine technical problem solvers. In fact, only a few of these people have the personal and technical qualifications for scientific inventiveness and prolific idea generating. A creative idea-generating scientist or engineer is a special kind of professional. This person needs to be singled out, cultivated, and managed in a special way. He or she is probably innovative, technically well educated, and enjoys working on advanced problems, often as a "loner".

The technical champion or entrepreneur is a special person, too. He or she shows creativity, but it is an aggressive form of creativity that is appropriate for selling an idea or product. The entrepreneur's drives may be less rational and more emotional than those of the creative scientist; he or she is committed to achieving but is less concerned about how to do so. This person is as likely to pick up and successfully champion someone else's original idea as to push something of his or her own creation. Such an entrepreneur may well have a broad range of interests and activities. He or she must be recruited, hired, managed, and stimulated very differently from the way an idea-generating scientist is treated in the organization.

The person who effectively performs project leading or project managing activities is yet a different kind of person. He or she is an organized individual, is sensitive to the needs of the several different people who are being co-ordinated, and is an effective planner. The ability to plan is especially important if long lead times, expensive materials, and major support are involved in the project development.

The information gate keeper is a communicative individual and is the exception to the truism that engineers do not read (especially that they do not read technical journals). Gate keepers provide links to sources of technical information which flows into and within a research and development organization and which can enhance new product development or process improvements. But those who do research and development need market information as well as technical information: What do customers seem to want? What are competitors providing? How might regulatory shifts affect the firm's present or contemplated products or processes? For answers to questions such as these, research and development organizations need people we call the "market gate keepers". Theses people are engineers, scientists, or possibly marketing people with technical backgrounds who focus on market-related information sources and communicate effectively to their technical colleagues. Such individuals read trade journals, talk to vendors, go to trade shows, and are sensitive to competitive information. Without them, many research and development projects and laboratories become misdirected with respect to market trends and needs.

Finally, the sponsor or coach is, in general, a more experienced, older project leader or former entrepreneur, who has a "softer touch" than when he or she was first in the organization. As a senior person, he or she can coach and help subordinates in the organizations and can speak on their behalf to top management. This activity makes it possible for ideas or programs to move forward in an effective, organized fashion. Many organizations totally ignore the sponsor role, yet our studies of industrial research and development suggest that many projects would not have been successful were it not for the subtle and often unrecognized assistance of such senior people acting in the role of sponsors. Indeed, organizations are most successful when chief engineers or laboratory directors naturally behave in a manner consistent with this sponsor role.

The significant point here is that the staffing needed for effective innovation in a technology-based organization is far broader than the typical research and development director usually has assumed. Our studies indicate that many ineffective technical organizations have failed to be innovative solely because one or more of these five quite different critical functions has been absent.

All of these roles can be fulfilled by people from multiple disciplines and departments. Obviously, technical people – scientists and engineers – might carry out any of the roles. But, marketing people also generate ideas for new and improved products, act as gate keepers for information of key importance to a project (especially about use, competition, and regulatory activities), champion ideas, sometimes sponsor projects, and sometimes even manage new projects, especially for new product development. Manufacturing people periodically fill similar crucial roles, as do general management personnel.

Multiple Roles

As indicated earlier, some individuals have the skills, breadth, inclination, and job opportunity to fulfil more than one critical function in an organization. Our data collection efforts with R&D staff show that a few clusters explain most of these cases of multiple role playing. Our common combination of roles is the pairing of gate keeping and idea generating. Idea-generating activity correlates, in general, with the frequency of person-to-person communication, especially with that which is external to the organization.[13] Moreover, the gate keeper, who is in contact with many sources of information, can often connect synergistically these bits into a new idea. This ability seems especially true of market gate keepers who can relate market relevance to technical opportunities.

Another role couplet is between entrepreneuring and idea generating. In studies of formation of new technical companies, the entrepreneur who pushed company formation and growth was found in half the cases also to have been the source of the new technical idea underlying the company.[14] Furthermore, in studies of M.I.T. faculty, 38 percent of those who had ideas that they perceived to be of commercial value also took strong entrepreneurial steps to exploit their ideas.[15] The idea generating-entrepreneuring pair accounts for slightly less than one-half of the entrepreneurs.

Entrepreneuring individuals often become project leaders. This progression is thought to be a logical organizational extension of the efforts of effectively "selling" the idea for the project. Some people who are strong at entrepreneuring also have the interpersonal and plan-oriented qualities needed for project leading. The responsibility for managing a project, though, is often mistakenly seen as a necessary reward for successful idea championing. This mistake arises from a lack of attention to the functional differences between the two roles. One should not necessarily assume that a good salesman will be a good manager. If the entrepreneur can be rewarded appropriately and more directly for his or her own function, many project failures caused by ineffective project managers might be avoided. Perhaps giving the entrepreneurs prominent project role. while clearly designating a different project manager, might be an acceptable compromise.

Finally, sponsoring occasionally evolves into a take-over of any or all of the other roles, even though it should be a unique role. Senior coaching can degenerate into idea domination, project ownership, and direction from the top. This confusion of roles can become extremely harmful to the entire organization: Who will bring another idea to the boss once he steals some junior's earlier concept? Even worse, who can intervene to stop the project once the boss is running amok with his new pet?

The performance of multiple roles can affect the minimum size group needed for attaining "critical mass" in an innovative effort. To achieve continuity of a project, from initial idea all the way through to successful commercialization, a project group must effectively fill all five critical roles. It must also satisfy the specific technical skill requirements for project problem solving. In a new, high-technology company, this critical mass may sometimes be ensured by as few as one or two co-founders. Similarly, an elite team – such as Cray's famed Control Data computer design group,

Kelly Johnson's "skunk works" at Lockheed, or McLean's Sidewinder missile organization in the Navy's China Lake R&D centre – may concentrate in a small number of select multiple role players the staff needed to accomplish major objectives. But, the more typical medium-to-large company had better not plan on finding, renaissance persons or superstars to fill its job requirements. Staffing assumptions should more likely rest on estimates that 70 percent of scientists and engineers will turn out to be routine problem solvers only, and that even most critical role players will be single dimensional in their unique contributions.

Career-Spanning Role Changes

We showed above some individuals fulfil multiple critical roles concurrently or in different stages of the same project. Even more people are likely to contribute critically but differently at different stages of their careers. This difference over time does not reflect changes of personality, although such changes do seem partly due to the dynamics of personal growth and development. The phenomenon also clearly reflects individual responses to differing organizational needs, constraints, and incentives.

For example, let's consider the hypothetical case of a bright, aggressive, young engineer who has just joined a company upon graduation. What roles can he play? Certainly, he can quickly become effective at solving routine technical problems and, hopefully, at generating novel ideas. But, even though he may know many university contacts and be familiar with the outside literature, he can't be an effective information gate keeper, for he doesn't yet know the people inside the company with whom he might communicate. Nor can he lead project activities, since no one would trust him in that role. He can't effectively serve as entrepreneur, as he has no credibility as champion for change. And, of course, sponsoring is out of the question. During the stage of his career, the limited legitimate role options may channel the young engineer's productive energies and reinforce his tendencies toward the output of creative ideas.

Alternatively, if he wants to offer and do more than the organization will allow, this high-potential young performer may feel rebuffed and frustrated. His perception of what he can expect from the job and, perhaps more importantly, what the job will expect from him, may become set in these first few months on the job. Though he may remain with the company, he will "turn off" in disappointment from his previously enthusiastic desire to make multidimensional contributions. More likely, he will leave the company in search of a more rewarding job. He will perhaps be destined to find continuing frustration in his next one or two encounters. For many young professionals, the job environment moves too slowly from the stage of encouragement of idea generating to a time when entrepreneuring is even permitted.

The engineer's role options may broaden after two or three years on the job, however. Though routine problem solving and idea generating are still appropriate, some information gate keeping may now also be possible as communication ties increase within the organization. Project leading may start to be seen as legitimate behaviour, particularly on small effort.[16] The young engineer's work behaviour may begin to reflect these new possibilities. Nevertheless, his attempts at entrepreneurial behaviour might still be seen as premature and sponsoring as still an irrelevant consideration.

After another few years at work, the role options are still wider. Routine problem solving, continued idea generating, broad-based gate keeping (even bridging to the market or to manufacturing), responsible project managing, and project championing may become reasonable alternative. Even coaching a new employee becomes a possibility. Though most people tend usually to focus on one of these roles (or on a specific multiple-role combination) during this mid-career period, the next several years can strengthen all these role options.

Losing touch with a rapidly changing technology may later narrow the available role alternatives as the person continues in his or her job. Technical problem-solving effectiveness may diminish in some cases, idea generating may slow down or stop, and technical information gate keeping may be reduced. Market or manufacturing gate keeping, however, may continue to improve with increased experience and outside contacts. Project managing capabilities may continue to grow as he or she tucks more projects under his or her belt. Entrepreneuring may be more generally sought and practised. This career phase is too often seen to be characterized by the problem of technical obsolescence, especially if the organization has a fixation on assessing engineer performance in terms of the narrow but traditional stereotypes of technical problem solving and idea generating. Channelling the engineer into a role that is more appropriate for an earlier stage in his or her career can be a source of mutual grief to both the organization and the individual. Such a role will be of little current interest and satisfaction to the more mature, broader, and now differently directed professional. An aware organization, thinking in terms of critical role differences, can instead recognize the self-selected branching in career paths that has occurred for the individual. Productive, technically trained people can carry out critical functions for their employers up to retirement if employers encourage the full diversity of vital roles.

At each stage of his or her evolving career, the individual can encounter severe conflicts between organization's expectations and his or her personal work preferences. This conflict is especially likely if the organization is inflexible in its perception of appropriate technical roles. In contrast, if both the organization and the individual are adaptable in seeking mutually satisfying job roles, the engineer can contribute continuously and significantly to innovation. As suggested in this illustrative case, in the course of a productive career in industry, the technical professional may begin as a technical problem solver, spend several years primarily as a creative idea generator, and add technical gate keeping to his or her repertoire while maintaining his or her roles. He or she may begin to serve as a project entrepreneur and lead projects forward. Gradually, he or she will develop greater market linking and project managing skills and eventually will assume senior sponsoring role, maintaining a position of project, program, or organizational leadership until requirement. This fully productive career would not be possible if the engineer were pushed to the side early as a technically obsolete contributor. The perspective taken here can lead to a very different approach to career development for professionals than is usually taken by industry or government.

Managing the Critical Functions for Enhanced Innovation

To increase organizational innovation, a number of steps can be taken to facilitate a balance of time and energy among the critical functions. These steps must be addressed explicitly or organizational focus will remain on the traditionally visible functions, such as problem solving, which produce primarily near-term incremental results. Indeed, the results-oriented reward systems of most organizations reinforce this short-run focus, causing the other, more significant activities to go unrecognized and unrewarded.[17]

Implementation of the results, language, and concepts of a critical functions perspective is outlined below for the selected organizational tasks of manpower planning, job design, and selection of measurement and rewards. If managers thought in critical functions terms, other tasks, not dealt with here, would also be carried out differently. These tasks include R&D strategy, organizational development, and program management.

Manpower Planning

The critical functions concept can be applied usefully to the recruiting, job assignment, and development or training activities within an organization. In recruiting, for example, an organization needs to identify not only the specific technical or managerial requirements of a job, but also the critical function activities that the job infers, e.g., the organization needs to ask whether the job requires that less experienced personnel be coached and developed in order to ensure the longer-run productivity of that area. If the job requires entrepreneuring, then the applicant who is more aggressive and has shown evidence of championing new ideas in the past should be preferred over the less aggressive applicant who has shown more narrowly technically oriented interests in the past.

Industry, at best, has taken a narrow view of manpower development alternatives for technical professionals. The "dual ladder" concept envisions an individual rising along either scientific or managerial steps. Attempted by many but with only limited success ever attained, the dual ladder reflects an oversimplification and distortion of the key roles needed in an R&D organization.[18] As a minimum, the critical function concept presents "multi-ladders" of possible organizational contribution; individuals can grow in any or all of the critical roles, while benefiting the organization. Depending on an organization's strategy and manpower needs, manpower development along each of the paths can and should be encouraged. Furthermore, there is room for individual growth and development from one function to another, as people are exposed to different managers, different environments, and jobs that require different activities.

Job Design and Objective Setting

Most job description and statements of objectives emphasize problem solving and sometimes project leading. Rarely job description and objectives take into account the dimensions of a job that are essential for the performance of the other critical functions. Yet, the availability of unstructured time in a job, for example, can influence the performance of several of the innovation functions, and it needs to be designed into corresponding jobs. To stimulate idea generating, some slack time is necessary so that employees can pursue their own ideas and explore new and interesting ways of doing things. For gate keeping to occur, slack time needs to be available for employees to communicate with colleagues and pass along information learned, both internal to and external to the organization. The coaching role also requires slack time, during which the "coach" can guide less experienced personnel.[19]

Essential activities for filling alternative roles also need to be included explicitly in a job's objectives. An important goal for a gate keeper, for example, should be to provide useful information to colleagues. A person who has the attitudes and skills to be to be an effective champion or entrepreneur could be given responsibility for recognizing good new ideas. This person might have the character to roam around the organization, talk with people about their ideas, and encourage their pursuit of these ideas. He could even pursue these ideas himself.[20]

Performance Measures and Rewards

We all tend to do those activities that will be rewarded. If personnel perceive that idea generating will not be recognized but that idea exploitation will, they may withhold their ideas from those who can exploit them. They may try to exploit ideas themselves, no matter how unequipped or uninterested they are in carrying out the exploitation activity.

For this reason, it is important to recognize the distinct contributions of each of the separate critical functions. Table 2 identifies some measures relevant for each function, indicating both quantity and quality dimensions. For example, an objective for a person who has the skills and information to be effective at gate keeping could be to help a number of people during the next twelve months. At the end of that time, his or her manager could survey the people who the gate keeper felt he or she had helped to assess the gate keeper's effectiveness in communicating key information. In each organization, the specific measures chosen will necessarily be different.

Table 2.   Measuring and Rewarding Critical Function Performance

Dimension of Management	Critical Function Idea Generating	Entrepreneuring or Championing	Project Leading	Gate keeping	Sponsoring or Coaching
Primary contribution of each function for appraisal of performance.	Quantity and quality of ideas generated.	Ideas picked up; percent carried through.	Project technical milestones accomplished; cost/schedule constraints met.	People helped; degree of help.	Success in developing staff; extent of assistance provided.
Appropriate rewards	Opportunities publish; recognition from professional peers through symposia, etc.	Visibility; publicity; further resources for project.	Bigger or more significant material signs of organizational status.	Travel budget; key "assists" acknowledged; increased autonomy and use for advice.	Increased autonomy; discretionary resources for support of others.

Rewarding an individual for the performance of a critical function makes the function more manageable and open to discussion. However, what is perceived as rewarding for one function may be seen as less rewarding, neutral, or even negative for another function because of the different personalities and needs of those filling the roles. Table 2 presents some rewards that seem appropriate for each function. Again, organizational and individual differences will generate variations in the rewards selected. Of course, the informal positive feedback of managers in their day-to-day contacts is a major source of motivation and recognition for any individual performing a critical innovation function, or any job for that matter.

Salary and bonus compensation are not included here, but not because they are unimportant to any of these people. Financial rewards should be employed as appropriate, but they do not seem to be linked explicitly to any one innovative function more than to another.

Performing a Critical Functions Assessment

The preceding sections demonstrate that the critical functions concept provides an important way to describe an organization's resources for effective innovation activity. To translate this concept into an applied tool, one needs to be able to assess the status of an R&D unit in terms of critical functions. Such an assessment potentially provides two important types of information: (1) inputs for management evaluations of the organization's ability to achieve goals and strategy; and (2) assistance to R&D managers and professionals in performance evaluation, career development, and more effective project performance.

Method of Approach

The methodology chosen for a critical functions assessment is contingent on the situation. From experience gained with a dozen companies and government agencies in North America, the authors have found the most flexible 5 to be a series of common questionnaires, which are developed from replicated academic research techniques on innovative contributors and modified as needed for the situation. Questionnaires are supplemented by a number of structured interviews or workshops. Data are collected and organized in a framework that represents: (a) the critical functions; (b) special characteristics of the organization's situation; (c) additional critical functions required in the specific organization; and (d) the climate for innovation provided by management. The results include a measure of an organization's current and potential strengths in each critical function; an evaluation of the compatibility of the organization's R&D strategy with these strengths; and a set of personnel development plans for both management and staff that support the organization's goals. This information is valuable for both the organization and the individual.[21]

Some Actions Taken in One Firm

As a result of a critical functions analysis in a company, multiple actions are usually taken. In order to consider some of the typical steps, we draw here from the outcomes implemented in one medium-sized R&D organization. The first action was that every first line supervisor and above, after some training, discussed with each employee the results of the employee's critical functions survey. (In other companies, employee anonymity has been preserved; data were returned only to the individual. In these companies, employees frequently have used the results to initiate discussions with their immediate supervisors regarding job fit and career development). The purpose of the discussion was twofold: to look for the differences in how the employee and his or her boss each perceived the employee's job skills; and to engage in developmental career planning. The vocabulary of the critical functions plus the tangible feedback gave the manager and the employee a meaningful, commonly shared basis for the discussion.

Several significant changes resulted from these discussions. A handful of the staff recognized the mismatch between their present jobs and skills. With the support of their managers, job modifications were made. Another type of mismatch that this process revealed was between the manager's perception of the employee's own perception. Most of the time the manager was underutilizing his or her human resources.

In this particular firm, the data also prompted action to improve the performance of the project leading function. An insufficient number of people saw themselves performing this function. Moreover, they saw themselves as lacking skills in this area. As a result of these deficiencies, upper management conducted several "coaching" sessions, worked to further clarify roles, and showed increased support for project leadership efforts.

Important changes also were made in how the technical organization recruited. The characteristic strengths behind each critical function were explicitly employed in identifying the skills necessary to do a particular job. This analysis led to a useful framework for interviewing candidates. It helped determine how the candidates might fir into and grow within the present organization. Upper management also became conscious of the unintended bias in the recruiting procedure. This bias was introduced both by the universities at which the company recruited and by the recruiters themselves. (In this case, the senior researchers, who conducted most of the interviews, were primarily interested in idea generating). As a result of the analyses, upper management was careful to have a mix of the critical functions represented by the people who interviewed job candidates.

The analyses led to other results that were less tangible than the above but equally important. Jobs were no longer defined solely in technical terms, i.e. in terms of required educational background or work experience. For example, if a job involved idea generation, the necessary skills and the typical activities for that critical function were included in the description of the job. Furthermore, the need for a new kind of teamwork developed since it was rare that any single person could perform effectively all five of these essential functions. Finally, the critical functions concept provided the framework for the selection of people and division of labour on the innovation team that became the nucleus for all new R&D programs.

Conclusion

We have examined the technology-based innovation process in terms of a set informal but critical behavioural functions. Five critical roles have been identified within the life cycle of activities in an R&D project. These roles are idea generating, entrepreneuring or championing, project leading, gate keeping, and sponsoring or coaching. In our surveys of numerous North American R&D and engineering organizations, we have made two key observations: some unique individuals are able to perform concurrently more than one of the critical roles; and patterns of roles for an individual often change over the course of his or her productive work career.

These critical functions concepts have managerial implications in such areas as manpower planning, job design, objective setting, and performance measurement and rewards. They provide a conceptual basis for design of a more effective multi-ladder system to replace many R&D organizations' ineffectual dual ladder system.

Several years of development, testing, and discussion of this critical functions perspective have also led to applications outside of R&D organizations. We have seen the perspective extended to such areas as computer software development and architectural firms. Recent discussions with colleagues suggest an obvious appropriateness for marketing organizations. A more difficult translation is expected in the areas of finance and manufacturing. To the extent that innovative outcome rather than routine production is the output sought, we have confidence that the critical functions approach will afford useful insights for organizational analysis and management.

References

1. For a different or more intensive quantitative view of project life cycles, see E.B. Roberts, *The Dynamics of Research and Development* (New York: Harper & Row, 1964).
2. See E. von Hippel, "Users as Innovators", *Technology Review*, January 1978, pp. 30-39.
3. For issues that need to be highlighted in a competitive technical review, see A.R. Fusfeld, "How to Put Technology into Corporate Planning", *Technology Review*, 80.
4. For further perspectives on project transfer, see E.B. Roberts, "Stimulating Technological Innovation: Organizational Approaches", *Research Management* (November 1979), pp. 26-30.
5. See D.C. Pelz and F.M. Andrews, *Scientists in Organizations* (New York: John Wiley & Sons, 1966).
6. See E.B. Roberts, "Entrepreneurship and Technology", *Research Management* (July 1968): 249-266.
7. See D.G. Marquis and I.M. Rubin, "Management Factors in Project Performance" (Cambridge, MA: M.I.T. Sloan School of Management, Working Paper, 1966).
8. See T.J. Allen, *Managing the Flow of Technology* (Cambridge, MA: The M.I.T. Press, 1977); R.G. Rhoaders et al., "A Correlation of R&D Laboratory Performance with Critical Functions Analysis", *R&D Management*, October 1978, pp. 13-17. Our empirical studies have pointed out three different types of gate keepers; (1) technical – relates well to the advancing world of science and technology; (2) marketing – senses and communicates information relating to customers, competitors, and environmental and regulatory changes affecting the marketplace; and (3) manufacturing – bridges the technical work with the special needs and conditions of the production organization. See Rhoades et al. (October 1978).
9. See Roberts (July 1968): 252.
10. One role we have observed frequently is the "quality controller" who stresses high work standards in projects. Other critical roles relate more to organizational growth than to innovation, e.g. the "effective trainer" who could absorb new engineers productively into the company, seen as critical to one firm that was growing 30 percent per year.
11. One study that demonstrated this phenomenon is N.R. Baker et al., "The Effects of Perceived Needs and Means on the Generation of Ideas for Industrial Research and Development Projects", *IEEE Transactions on Engineering Management*, EM-14 (1967): 156-165.
12. Section VI describes a methodology for collecting these data.
13. See Allen (1977).
14. See Roberts (July 1968).

15. See E.B. Roberts and D.H. Peters, "Commercial Innovations from University Faculty", *Research Policy*, in press.

16. One study showed that engineers who eventually became managers of large projects began supervisory experiences within an average of 4.5 years after receiving their B.S. degrees. See I.M. Rubin and W. Seelig, "Experience as a Factor in the Selection and Performance of Project Managers", *IEEE Transactions on Engineering Management*, EM-14 (1967): 131-135.

17. For further perspectives on the consequences of this short-run view by U.S. managers, see R.H. Hayes and W.J. Abernathy, "Managing Our Way to Economic Decline", *Harvard Business Review*, July-August 1980, pp. 67-77.

18. For a variety of industrial approaches to the dual ladder, see the special July 1977 issue of *Research Management* or, more recently, *Research Management*, November 1979, pp. 8-11.

19. In a more macroscopic way, March and Simon observed years ago that innovation could only occur in the presence of organizational slack. See J.G. March and H.A. Simon, *Organizations* (New York: John Wiley & Sons, 1958).

20. For more details on various job design dimensions appropriate to the critical functions, see E.B. Roberts and A.R. Fusfeld, "Critical Functions: Needed Roles in the Innovation Process", in *Career Issues in Human Resource Management*, ed. R. Katz (Englewood Cliffs, NJ: Prentice-Hall, forthcoming).

21. For samples of questionnaire items, more details on diagnostic uses of the resulting data, and numerical outputs from one company's assessment, see Roberts and Fusfeld (forthcoming).

Case Study on Creativity and Innovation: The Approach of ENEA and the "Prato Case"

Raffaele De Maria, ENEA, Bologna, Italy

As a start, I am going to refer to the ENEA intervention strategy to support PMI, particularly in the industrial district; then I shall examine more specifically the activity carried out in the Prato Textile District.

I would like to point out that I do not mean to present all the activities being carried out or planned, but rather a synthesis of the main guidelines of what has been experimented, the problems faced and the aims reached.

Elements of the ENEA Strategy

Some Remarks on the Productive SME System

As everybody knows, a productive system mainly based on small and medium sized enterprises (SMEs) and handicraft is one of the most important characteristics of our country.

For instance, handicraft enterprises are considered to be about 1,400,000, more or less 38% of the EC craftsmen (for a comparison 6% for France and 8% for Germany).

Italy is not the only example of a large SME presence; Japan is another example. But while in Japan the development has concentrated towards high technology industry, in our country dynamism and innovation have developed starting from the adaptation and use of the new technologies to support the development and competitiveness of traditional sectors such as textiles, wood, leather and mechanics.

This has brought about a considerable growth, in particular in the regions of the North-East and the Centre of Italy.

In spite of the apparent fragmentation, this growth has been helped by wide and strong social relations and entrepreneurial capability, which ensure a flexible specialization and can meet an often variable market demand.

We are referring to the reality of the Italian industrial district. The industrial district is essentially the consequence of historical process; it is founded on a specific ethic that you can try to define as the ethic of sacrifice, of work, of family. In this context the role of the family has been fundamental. Very often you have a sort of identification of the family group with the firm.

The existence of some kind of combination of shared values and the persistence of a system of value is a preliminary requirement for the beginning and the development of the district since among others it ensures the possibility of creating trust, the disposition to exchange information and loyalty.

In this framework a very important role in order to determine the existence of a stable community has been played by the local parties and the local governments.

In the district the job is characterized by a combination of job positions: home – based work, part-time, waged work, self-employment, entrepreneurs. In a growing district the ability is better appreciated than in other industrial situations, there are links between the productive activity and daily life, an important part of the family income is to be ascribed to a condition near full employment.

Moreover the social politics and a system of infrastructures have favoured a high growth of women's employment.

In a district you have a very high level of information flows. The distinguishing fact is that these are usually of the face to face type. The existence of this person to person contact is also essential for a high level of creativity, to find solutions and convert quickly innovative ideas into products on the market. In fact in the district it is easy to share information belonging to different functional and technological areas; and this creates information flows essential to creativity and diffusion of innovation.

From another point of view, through the peculiar information flows in the district, you manage to produce a specific technological culture and those who belong to a district consider being technologically up-to-date as a measure for a better future.

The existence of a local bank, closely linked to local firms and institutions, which knows the life of the district and is able to give the right weight to personal qualities, is very important. At the same time a good management of the local bank in the district is essential; otherwise, due to the interdependence typical of the district, you can have a chain of negative effects impairing also the system of value of the district. Often representatives of the local institutions are members of the board of directors of the local bank.

Local institutions exert a very important role through the management of different factors, such as the function attributed to the local government and the transmission of technological knowledge by means of specialist technical schools.

It seems moreover relevant to emphasize the role of associations among producers and other local institutions with respect to the establish of prices.

You can say that these prices are local prices in the sense that, even if it is unavoidable to take into account results from the national and international market, at the same time you cannot forget the local conditions of demand and supply.

So the productive organization typical of the districts and the very wide presence of subcontractors have important consequences on the establishing of labour costs.

The associations and also the local institutions exert a very important role of stabilization and regulation, thus contributing to the trust and the loyalty in the district.

In general terms you can say that you need the presence of a recognized association system in order to favour the existence of efficacious conditions of co-operation / competition.

Some General Elements of ENEA Intervention Methodology

In this reference frame of the Italian Industrial system, the strategy of a public "operator" such as ENEA takes place; today ENEA institutional task is in particular devoted to the production and diffusion of technologies.

In the last ten years ENEA has consolidated a programme in favour of SMEs and handicrafts and therefore of the industrial districts, thinking that it is wrong to consider some sector as mature, if what is meant by this word is an established technology. In short, ENEA intervention consists in:

> acting as a technological agency with interventions strongly connected to the socio-economic territorial aspect, and therefore not like an external technology supplier who does not know the customer's needs, but like an operator who can meet them;

supplying real services, not just financial incentives; these services will involve enterprises both in finding out the needs and in providing the facilities to supply the services;

supporting a production system as a whole (and therefore not just advising a single firm), encouraging collaboration in order to find solutions to common problems through joint efforts and common resources.

At the same time ENEA will favour collaboration among suppliers of new technologies, such as Universities, advanced industries, ENEA laboratories and the users of the innovations, in particular, SMEs and craftsmen.

On the basis of these considerations, the experience of these years has led to an intervention methodology in industrial districts such as the one of Prato.

In spite of its changing according to the different needs of the various districts, this methodology has common elements that can be summarized as follows:

Necessity to create a confrontation – Meeting basis in order to represent the socio-economic system in the best way and to guarantee the quality of the intervention.

This reference point should be a basis to have always a strong potential of common interests.

Need to create the conditions for a financial participation in the intervention of the different subjects, this participation being also a necessary guarantee for its success.

Choice of institutional forms of support of the agreements reached, based on existing forms of collaboration, on association structures, on local companies involved in the plans.

On the basis of these elements the experiences carried out up to now have, in short, led to form a group representing the various partners (this group could be generically called "Consiglio Direttivo") and a technical structure working as an instrument both for organizing intervention suggestions and carrying them out.

Work according to this methodology has stressed some difficulties that can be summarized as follows:

The modernization and the innovation diffusion in the SME and the handicraft systems, mainly concentrated in "traditional" productive processes, are rather complex to plan, since it is necessary to find the best possible meeting point between a socio-economic and cultural context (broadly speaking) of a district or a sector included in a district, the available financing resources and the technology "market";

The need to guarantee a continuous joint intervention involving more partners gives ENEA not only the task of evaluating, supporting and guaranteeing the technological elements but also the role of an operator above the parties. The presence of such an operator is fundamental;

The dynamic evolution of the areas involved and the time-factor are other elements difficult to be dealt with and require a capability of continuous monitoring, the use of new instruments, strategic revision; this is true also in the case of the institutions organization system;

The analysis of the cost–effective aspects of the interventions which are financially comparatively "modest", based more on the quality of the expense and the availability of competent people at high professional level than on the quantity, is not easy, just because of the complexity of the actions to be carried out and the different characteristics of the situations and structures involved. It requires a definition and application of specific methodologies.

The Sprint Prato Intervention

A Few General Elements

The Prato textile system is characterized by the presence of about 12,000 firms (80% handicraft), about 50,000 workers, a billing of about 5,000 billions Lire, almost half of which comes from the exporting of the textile products.

The system is mainly based on entrepreneur called "impannatore", whose role is to acquire orders and to act as an intermediary with the market, to organize the production and distribute the tasks among the different firms (mainly small firms) and the craftsmen, specialized in single production phases.

In this complex system firstly ENEA has promoted the establishment of an Association named SPRINT (System Prato Technological Innovation) of which partners, beside ENEA, are the Entrepreneurs Association of Prato, Craftsmen Guilds, Local Banks, Chamber of Trade, Textile Trade Union, STET (State owned Telecommunications holding), Municipality of Prato, secondly ENEA has taken part actively in the design, realization and financing (about 45%) of the programmes of SPRINT.

At present SPRINT is carrying out a set of projects aiming at ameliorating the global competitiveness of the area and at supporting the evolution of the socio-economic context. The most important topics are:

The communication system and telematics;
The quality and the promotion of quality services for SMEs;
The diffusion of advanced management systems specifically developed for the firms of the area;
The support of the diversification and in particular of the technical textiles;
The market and business monitoring;
The diffusion of the technical information in particular by the Newsletter and the SPRINT technical papers.

The Telematic Project

The SPRINT Telematic Project is the most advanced example of the model carried out by ENEA to support the development of the telematic services in a district area such as Prato as well as for a collective framework such as a municipality or a public services supplier.

The most important choices of the model can be summarized as follows:

The success must rely more on the efficacy of the services than on a technological angle. So we use the normal telephone network, low cost hardware for the interconnections as well as for the computers;

The network of the services rely on a pre-existing system of commercial and productive or social relationships;

An accurate analysis of the communication fluxes has to be carried out;

The services must offer profit opportunities in a framework that has to ensure reliability, technical assistance and quality;

A telematic centre has to be provided. It has to operate as a pole able to link and to integrate and to produce goods and services in the communication field. In particular, you have to connect the local and territorial needs and resources with the services specifically developed for the district or available at a more general level.

This approach has been tested in particular in the Prato textile district where the analysis of the very complex communication system has proved that to ameliorate the information system of the area could determine important advantages for the competitiveness of the district.

In brief the project has been carried out as follows:

The video text system has been adopted;
The most important social and economic actors of the area have taken part in the Project, operating as information supplier too;
An experimental phase has been carried out in accordance with SIP and STET with the participation of about 450 subscribers. This phase was concluded at the end of 1990.

Now we are engaged in a more commercial phase aiming to the capillary diffusion of the services, through an agreement with a private firm (PRISMA) too.

The Quality Project

The district of Prato has to face important quality problems due to both the need to confront more and more difficult trade competition and the quality standards requested in future, such as the rules arranged by EEC.

In brief the principal aims of the Quality Project are:

The study of the quality problems of the textile and clothing sector and the application of the results in the Prato district;

The diffusion of the quality culture in the area and vocational training of entrepreneurs, technicians and managers;

The design, the planning and the promotion of a Centre devoted to place specific services at a firm's disposal.

This Textile Quality Centre has been operating since 1991 and SPRINT is member of the board of directors.

Management Advanced System Project

The Management Control Project was introduced in order to solve problems proposed by some of the most representative firms of the District aiming at reducing costs and at ameliorating the relations with the market.

The Project has been carried out in two phases. During the first a general model has been developed taking into account the specific characteristics of the textile business.

Subsequently a complete firm model has been designed, which takes into account the different firm areas – administration, finance, trade, production, and logistics.

The firms that are using the model have introduced important alterations in the structure of the organization and in the information system.

This Project has shown that it is possible to introduce advanced management system in the SMEs firms and to optimize costs and planning in the textile sector.

Environment Project

Industrial districts have a peculiar situation with regard to environmental questions due to the very high number of SME firms operating in a restricted territory. In fact, the district represents a specific "ecosystem" and when you treat the environmental problems you have to take into account a lot of factors characterizing the district such as territorial characteristics, institutions, productive typology, environmental emergency, population, as well as the role performed by the system of relations peculiar to this particular district.

The research work of SPRINT aims at introducing a systems point of view in a district, in which you have already adopted important measures in order to solve specific environmental problems while a more global approach is not yet usual.

At the same time you are helping to face recurrent difficulties concerning water pollution and shortage, you are testing non-polluting processing techniques, such as laser surface treatments, you are studying the reuse of textile wastes.

Technical Textile

The technical textile market is growing at a much faster rate than other textile sectors in the EC and it is characterized by a very great number of possible applications.

SPRINT aims at supporting the development of the technical textiles in the district because it is important to favour the introduction of new processes, not linked to the fashion system, today largely prevailing in the area.

Technical textiles can represent a filler parallel to the traditional production of the district, more founded on the research, development and technology and connected with markets different from that related to the fashion.

This project, started last year, is working on market assessment, in order to point out the most promising fields for the firms of the district, on the vocational training, that is very important in the case of the technical textiles, on the set up of a cluster of firms involved in the research and development.

Market Monitoring and Information Diffusion

In 1987 the Entrepreneurs Association stimulated the development of a market survey that would provide regular information on the status of the textile industry of the district.

SPRINT has been charged to realize this service, due to the necessary requirements of independence and collective nature ensured by SPRINT.
A permanent panel of about 400 firms, representative of the textile productive cycle of the district, has been activated and a quarterly analysis of the economic trend is performed.

Annually, partially based on the same panel, more general information on the situation and the evolution of the area is released.

With reference to the diffusion activities SPRINT publishes a monthly Newsletter and a quarterly technical paper on technological subjects significant for the area.

Problems and Suggestions

The problems to be solved in order to implement the model are of different type:

Cultural. In some respect in our country there is a substantial delay as regards to the diffusion of the innovation and this has as a consequence that you have to make an important effort in order to ensure the diffusion of a specific culture and technology in the area where you are operating;

Economic. You need to find the resources in order to ensure the start-up of an action for which it is difficult to foresee an immediate profit. But this is not easily accepted by SMEs;

Strategic. You have to cope with a matter like innovation, which is connected to a very large number of different factors, not only technology but also market, competitiveness, etc. These factors are difficult to manage at the same time and in particular in a very fragmented productive cycle;

Technical. For instance at the beginning the Videotel system was not so reliable; but reliability is a very important requirement in particular when operating with SMEs;

Planning. It is necessary to ensure a continuous "tuning" of the projects in order to have an adequate participation of all the actors in the initiative.

In brief we can get the following suggestions from our experience:

You need a strong consensus about this type of intervention;
You have to organize and develop local abilities;
The time factor is very important and you have to take into account the particular financial and organizational characteristics of SMEs;
You have to support the development of specific co-operation agreements;
You need specific tools for the monitoring and the evaluation of the intervention.

Establishment of the Modern Patent. Legislation System in Russia

Vitaly P. Rassokhin, **Committee for Patents and Trademarks of the Russian Federation, Moscow, Russia**

The Patent Law of the Russian Federation came into effect on October 14, 1992. Its introduction in Russia removed the legal uncertainty about granting patents that had existed since the former USSR had collapsed. Russia, now, has a European type law that meets all the requirements for the harmonization of patent legislation.

The main difference that the new law has over previous acts in the field of industrial property protection, is that it covers items that had not been protected earlier. Article I of the Law reads that this Law governs relations arising in connection with the creation, legal protection and exploitation of inventions, utility models and industrial designs. Utility models are a new object for protection in Russia.

Those involved in developing the new Russian patent legislation justified including patent protection for utility models (technical solutions relating to structures but not having the level of inventive step for recognizing them as inventions) for a number of reasons. First, this object is present in the Paris Convention for the Protection of Industrial Property (to which Russia as the successor of the USSR is a member), and it is also legally protected in 30 countries. A number which is growing regularly. Second, most innovative activities are now undertaken by small and medium sized enterprises which need cheap, operative and sufficiently effective protection of their developments. The inclusion of utility models complies with these requirements to a greater degree. Third, the statistics on patents granted for utility models abroad confirm that they are created, mainly in branches of industry producing consumer goods, which is particularly important for Russia now.

Another peculiarity of the Patent Law is the absence of norms in it relating to government management over inventive activities and the use of inventions in the national economy which were a common feature of the USSR Patent Law.

The draft Patent Law of Russia treats such important problems as "employees' inventions" (or other patentable objects created in the course of performing regular work) in quite a different way than it was used in the

Law of the USSR. The interests of the investor, i.e., employer, are put to the forefront. Therefore, the primary right to obtain a patent for an object created by an employee in performing a specific or assigned task is secured to the employer.

Taking into account the right to commercial secrecy attributed to the employer by the corresponding Russian laws, the Patent Law allows the employer to avail himself of filing an application in cases where a specific object should be kept secret. In such a case an inventor does not acquire the right to obtain patent in his name, but he keeps the right to receive a remuneration if the unpatented solution is exploited. The Law protects the interests of the inventor also in the case where the employer has filed an application for a patent, but has not obtained it due to certain actions (e.g., he might withdraw his application or fail to pay the prescribed fees or avail himself of the patent already granted). In each such case the inventor is entitled to demand that compensation be paid to him by the employer, since he has been deprived of the possibility of obtaining a patent in his own name.

Where a patent for the employee's development is granted to the inventor (under a contract concluded between him and the employer or where the latter is not interested in obtaining a patent), the employer, under the Law, is entitled to use it in his production without signing a licensing contract, but subject to payment of a reasonable compensation to the patent owner.

The Russian Law reiterates the criteria of patentability established earlier by the USSR laws on inventions and on industrial designs, as well as the usual list of non patentable subjects matter. In this connection it should be mentioned that the legislation took into account that the criteria of patentability set forth in the laws of the USSR are harmonized with the laws applied at present in most industrially developed countries.

As far as utility models are concerned, their list is confined to such a category as devices (structures): they are not required to possess an inventive step, but changes in the prototype should affect its essential features and, respectively, influence the technical results achieved while using such an object. For utility models the list of materials considered in the course of assessing patentability is limited, and includes only information on objects for the utility model's claimed purpose.

The procedure of granting patents laid down in the Russian Patent Law is also different from that fixed in the USSR Law on Inventions. It provides for the system of deferred examination which is characterized by the fact that the examination is to be carried out at the special request by the applicant, rather than automatically. The Law provides a 3-year term from the filing date of the application with the Patent Office for filing of uncertainty with respect to legal protection of the claimed invention, affecting the interests of third parties, the latter are also given the possibility of filing requests for examination during the said period. Moreover, both the applicant and any third party may request that a state-of-the-art search be conducted with regard to the application for the purpose of assessing the novelty and the inventive step of the claimed solution. Having the results of such a search, the applicant may assess the possibility of obtaining a patent and the expediency of filing a request for examination, and third parties may consider the "risk level" of the filed application from the point of competition.

With respect to utility models the Law provides another system of granting patents called the "registration system". A patent may be granted for any duly executed application and for any solution relating to the category of objects protectable as utility models. The decision whether the utility model is patentable or not will be checked only in a case where the patent granted is challenged. This system is commonly adopted (Japan is the only exception) for utility models.

The patent of the Russian Federation does not have any legal features differing it from a classical patent. Thus its nature consists in the exclusive right of the patent owner to exploit inventions, utility models and industrial designs protected by patents in the territory of Russia. This right includes the right to prevent other persons from using patented objects.

Since the State is interested in the exploitation of patents primarily for the purpose of creating favourable conditions for entering the commodity market by the patent owner, and not for the purpose of blocking entry by his competitors, the Law provides for the responsibility of the patent owner to exploit the protected object in the territory of Russia. In the case where the patent owner fails to exploit or insufficiently (not satisfying the market demand) exploits the patented object during four years from the date of granting of the patent for invention or industrial design or during three years from the date of granting of the patent for utility model, a compulsory license may be granted if a person who is ready to exploit the protected object wishes so. The point is that, if the patent owner fails to exploit the patented object and, at the same time, refuses voluntary licenses, any interested party may lodge a petition to the Patent Court of the Russian Federation requesting a compulsory license. The Patent Court grants a compulsory license, at the same time deciding on disputed questions on limits of exploitation, amounts, time limits and procedure of payments which may not be lower than the market price of such a license.

The above Patent Court (in the Law it is called as a Higher Patent Chamber) has as its main function to take final decisions on patentability of claimed objects as well as opposition to patents granted for inventions or industrial designs. The complete list of disputes that will be under the jurisdiction of the Patent Court is supposed to be contained in a special law on the Patent Court of the Russian Federation.

Let me dwell upon two more peculiarities of the Russian Patent Law.

It does not provide for the permissive manner of patenting inventions, utility models and industrial designs in foreign countries. Obligatory is only prior filing of applications by individuals and legal entities of the Russian Federation with the Patent Office of Russia.

On secret inventions, the Russian Patent Law provides the condition that the patent form of protection must not be used in respect of inventions and other objects of information which are not accessible.

At present, the field of industrial property in the Russian Federation is in a state of transition. In this period applications for inventions, industrial designs, utility models and trademarks are accepted, the said industrial property objects are examined for compliance with the protectability criteria, and corresponding official actions are prepared. After the Rules and Procedures of the Russian Patent Law are put into effect, patents will be granted for such applications.

The applicants that have filed their applications for USSR inventors' certificates or patents for inventions or industrial designs, if such applications are still pending and titles of protection are still not granted by the effective date of the Patent Law of the Russian Federation, are given the right to apply for patents of the Russian Federation, while keeping the priority of their initial applications. Requests to that extent should be filed with Rospatent – the Patent Office of the Russian Federation.

Such applications will be considered in the manner stipulated in the Patent Law of the Russian Federation. The conditions of protectability of the invention or industrial design, that were set forth in the legislation valid at the filing date of the application, will be applied.

It is important to note that the validity of USSR titles of protection issued earlier for invention and industrial designs is recognized in the territory of the Russian Federation including inventors' certificates. In accordance with the Resolution of the Supreme Soviet of the Russian Federation on Putting into the Effect the Patent Law, inventors' certificates of the USSR (as well as certificates on industrial designs) may be (upon requests of inventors and applicants) transformed into patents effective in the territory of Russia.

Thus, on the basis of such a request the right of the state is terminated and the exclusive right of a definite person – usually of an inventor himself – arises, i.e. in other words "privatization of intellectual property" takes place.

The clear legal protection of such objets as trade marks and appellations of origin of goods will foster the development of the market economy in Russia.

The growing interest towards the acquisition of the rights for trade marks is proved by statistics of the quantity of applications for their registration for the last few years. For example, in 1980-88 the number of such applications was approximately from 5 to 7 thousand a year, in 1989 their number was 12,000, in 1990 – 20,000, in 1991 – 26,000, in the first half of 1992 – 15,000.

On October, 17, 1992 the Law "On Trademarks, Service Marks and Appellations of Origin of Goods" came into force in the territory of Russia. The Law has a number of differences from the legal base that existed before in Russia and the USSR in this area.
All the normative acts adopted earlier in this area were governmental or ministerial acts. Upon putting the Law into effect, legal relations arising in connection with the mentioned objects will be regulated at the legislative level.

Legal equality among trade marks and service marks is a distinctive feature of the Law. This can be proved by the mentioning of "service marks" in the title of the Law.

Another peculiarity of the Law is that it regulates legal relations connecting such an independent object of industrial property as Appellations of Origin of Goods.

The legal protection of this object relating to goods and combining in itself the core of material and spiritual cultures acquires social importance especially for Russia which is a multinational state. The protection of appellations of origin of goods will promote the preservation and development of the traditional handicraft and domestic industries reflecting the unique history and culture of people of different nationalities who populate the Russian Federation. It should also be noted that the Law which introduces the legal protection of appellations of origin of goods is adopted for the first time in this country.

The important Law of the Russian Federation, the Resolution of the Supreme Soviet of Russia on putting into effect the provisions of Civil and Criminal laws which concern the above objects directly, all establish a necessary legal mechanism for the protection of rightholders' trademarks and registered users of appellations of origin of goods as well as for the elimination of unfair competition and thus they will secure the consumer interests. For the latter the above indications are the means of the better orientation when selecting the proper goods.

The introduction of the clauses into the Law stipulating measures of responsibility for unlawful use of trademarks and appellations of origin of goods is a quite timely move as there are cases of the production of counterfeited goods on the territory of Russia with the use of trade marks belonging to famous foreign companies.

Conceptual provisions of the Law are in full conformity with the international conventions, to which the Russian Federation is a party, and they are harmonized with the laws of the leading countries of the world.

According to the Decree of President Yeltsin the functions of the patent office of Russia are fulfilled by the Committee of the Russian Federation for Patents and Trademarks which I have the honour to chair. This Committee is an independent state body subordinated directly to the Government of the Russian Federation.

In accordance with the above-mentioned Laws on the protection of industrial property foreign applicants are to file their applications in Russia through the patent attorneys registered with the Committee of the Russian Federation for Patents and Trademarks. At present those patent attorneys who have been registered with the former USSR Gospatent, are still recognized. However, after the adoption of the statute on patents, the official registration of each patent attorney with the Committee for Patents and Trademarks will be necessary.

The Transition to Market Economy and Innovation Management

Boris Z. Milner, Institute of Economics, Russian Academy of Sciences, Moscow, Russia

The transition of Russian economy to market forms, being held under the conditions of sharp political struggle, deep economic crisis and the growth of social tension, is coming to its decisive stage. The main directions of the reform are:

Deregulation of the economy;
Finance and money stabilization;
Privatization;
Active social policy;
Structural reconstruction of the economy;
The creation of subject of market economy and competitive environment.

The liberalization of prices was introduced in early 1992. Tough financial policy, directed towards balancing incomes and revenues of the population, is being fulfilled. Nationalized property is being privatized. Administrative restrictions on economic ties and external economic links are being removed. Agriculture reform, aimed at creating a vast majority of farmers has started. A market infrastructure, that is commercial banks, commodities and stock exchanges, brokers offices, commercial firms, contracts system, is being formed.

Speaking about the first achievements of unprecedented economic reform in Russia, it can be noted that 10 per cent of the population is already in some aspects connected with entrepreneurship. 24,000 enterprises in different branches of economy have been privatized during the first 9 months of 1992. More than 4/5 of all the privatized enterprises are in the spheres of commerce, public catering and services.

Large and middle-scale state enterprises are being transformed into joint-stock companies – to the end of 1992 five – six thousand of the enterprises will become joint-stock companies. Vouchers are being distributed among the people. By October, 1 1992 about 150,000 farms have been created. About 1,300 enterprises in the defence sector and 920 research institutions have taken part in conversion programmes. The volume of military production in 1992 has been 4 times less than in 1988.

But these and some other measures, aimed at the Russian economy's transformation, have not given tangible results yet: production is declining, inflation grows, investment activity and the population's revenues decline, employment is reduced. Since 1992 the economic situation in the country has deteriorated and has possessed some new features.

First of all, the production decline became overall and reached volumes leading to physical distortion of productive forces, the destruction of production, scientific and technical potential. There is no possibility to slow down the production decline this year. During the last months it has even become faster. As a whole during the first 9 months of 1992 the volume of production has been reduced by 17,6 per cent, and in July, August and September – 22, 27 and 25 per cent correspondingly.

The sharp reduction is typical for the regions with high concentration of the enterprises under conversion, limited possibilities of resource self-provision, and social and political tension.

It should be stressed, that there are tendencies, which make inevitable further reduction of production in the near 1,5 – 2 years. Among them there is a sharp reduction of investment activity. Investments of enterprises and organizations of all forms of property are 40 per cent less than during the first 9 months of the last year. Decentralized investment held by enterprises has reduced by 43 per cent.

Analysis shows that crisis is connected with limitation of state investment resources and also with inflation, which prompts the enterprises to cut down long-range construction programmes.

It should be added, that hyperinflation has turned out to be a real factor. Consumer prices for commodities and services grew 13,1 times in September 1992 as compared to December 1991.

Deep causes of the crisis lie in long-lasting occupation of administrative system and caused by its degradation of all economic and social structures. It is this that caused the depreciation of equipment, sharpening of resource disbalances in foods, raw materials, fuel and energy, the reduction of fuel resources mining and reduction of revenues from external economic activity. These causes were added to by new ones, connected with objective difficulties of the transition period from administrative to market economy, with an unstable political situation, with power crisis and multinational conflicts. In the light of peculiarities and crises, phenomenon of the transition period from centralized management to market relations in the economy, the innovation crisis becomes apparent very clearly. The main contradiction, which causes this crisis, is that, from one side, the transition to free entrepreneurship, the creation of competitive environment, orientation on consumer needs must stimulate initiative activity, technological innovation, realization of scientific and technical innovations. From the other side, deep economic crisis, accompanying the transition period, sharp reduction of investment activity, social tension in the society disorganize the innovation process, liquidate the natural incentives to innovation activity.

In these conditions the innovation crisis is realized in a sharp decline of managerial level of the process of innovation's creation and realization, in the absence of the source of its financing, in reduction of research and scientific groups. A severe blow on innovation activity is the "brain drain" to other countries.

It should be stressed, that one of the causes of innovation potential reduction is the lessening of state expenditure on research and development, caused by the economic crisis. It has already caused the reduction of financial support to fundamental research, a number of scientific institutions have been closed, engineering and technological centres and branch funds for scientific support have been liquidated. For example, due to the absence of money resources the institutes of the Academy of Sciences have practically stopped buying equipment and reduced current expenditures on scientific work. The suggestion of liquidating some scientific institutions is under discussion. Some of the departments will be converted to other institutions with common profile. A number of organizations will have to be deployed on smaller rooms in order to reduce rent payment. According to the estimations, the Academy's budget deficit in the last quarter of 1992 may reach 1 billion roubles. We expect, that in the near future, due to the corrections of economic reforms course and the adoption of a number of additional anti-crisis measures, negative tendencies will be overcome, production activity will be renewed, hyperinflation will be overcome, and the managerial capacity of national economy will be restored. An important role is given to the use of high technology in the defence complex for the production of competitive foods, and the stimulation of innovation activity. Speeding the innovation activity development at a new stage of reform, as we see it, will be supported by the measures to support entrepreneurship. The main thing is to create favourable conditions for fast growth of private sector in the economy, to create conditions for including entrepreneurs into the market relations.

For this purpose the programme of new stage of the reform includes:

Giving privileged credit to new entrepreneurs for 2 years;

Privileged access of private entrepreneurs to foreign credits, given by state or by international economic organizations, through competing investment and some other projects;
Privileged (50 per cent) taxation of reinvested profit of private enterprises;
Creation of the Fund of entrepreneurship support;
Guaranteeing stable legislation on the protection of entrepreneurship.

We expect that privatization and development of the entrepreneurship in the next 2 or 3 years will cause the growth of business and innovative activity in Russia. High level of privatization will guarantee a radical change of economic situation.

In the near future structural reconstruction of the Russian economy, based on demilitarisation and adaptation to the structure of consumers demands, will play a significant role in the process of transition to market economy. The main priorities of structural policy at present are fuel and energy, foods, overcoming of the housing crisis and communications. The course should be taken on social orientation of the economy and modernization of its industrial apparatus, the speeding of resource-saving technologies, outstripping growth of scientific production export. It has been realized that these priorities will be based not on the mass state investments or on centralized distribution of material resources, but on the methods of economic regulation, activization of entrepreneurship, growth of economic activity of the population, attraction of foreign capital and technical experience.

For example, demilitarisation of the industrial productive sector and of R&D is planned to be held on the basis of non-state forms of property and entrepreneurship, reduction of state subsidies and curtailment of ineffective industries and enterprises and removal of limitations of foreign competition.

The main directions of the innovation process development in Russia, in accordance with economic reforms being fulfilled, are as follows:

1. As decentralization, demonopolization and privatization of state property are realized, the efforts to develop science and technology, innovation should be concentrated on the level of economic units that is enterprises, scientific and technical centres, associations, technoparks, and other scientific and industrial institutions.
   It is possible to use in this connection the mechanism of the stock exchange to master scientific and technical achievements, which

will accelerate the innovation process. The activity of the above mentioned institutions must orient mainly on applied research works, aimed at the satisfaction of consumer market demands in the near future and at the elimination of existing deficits. The adaptation of the R&D sphere to market conditions is expected to be difficult and lasting. The adaptation difficulties are caused by the absence of the needed infrastructure, institutional, legislative and economic environment for independent innovation activity of economic units in a new situation.

2. The system of state regulation of the R&D sphere and fundamental research should be organized in a new way, large-scale applied inter-branch projects, all Russian programmes (ecology, medicine, education, training and so on) are possible to be fulfilled only at the expense of massive budget grants. The creation of a favourable innovation climate, the determination of strategic directions of scientific and technical development and the participation in international technological exchange should become the important constituent parts of state scientific and technical policy.

3. The formation of the system of legislative regulation in the sphere of R&D as applied to market economy's conditions. Such a system should include such legislative acts:

   About scientific and technical policy;
   About the status of scientific establishments and scientists;
   About the protection of inventions and some other objects of intellectual property;
   About privileged taxation of enterprises, producing new technology;
   About the mechanism of credits to enterprises and price-marking for scientific products with priority usage and so on.

It is quite evident that we face the necessity to create legislative, institutional and economic instruments, which have been absent in the administrative and command system and is vitally important for including the innovation potential of Russia into its national economy, into wide international ties.

It is in Russia, where the role of state property is large, and so the optimal innovation system should combine state support to innovations introduction and concrete introductive activity of non-state institutions. Such a combination is much more effective and dynamic, that "pure" state organization of scientific and technical progress and managing the shifts in national economy's structure.

The Republican Innovation Fund has been created for financial support to innovation programmes. Its non-governmental means account for 50 per cent in order to combine state and entrepreneurs interests. The Russian government delegated to Investment Fund the rights of state representative, its relations with the state are being built on a contractual basis. Credits for all-Russian and regional programmes are given on the security of federal and municipal state property. Priorities in the elaboration of republican innovation programmes are given to such spheres and branches, as foods, medicine, construction materials and technologies, consumer foods and ecology.

The prospects of essential activization of innovative activity should be closely connected with wide participation of foreign capital and technology in the development of Russian industrial, scientific and technical potential. It should be taken into account in this connection, that close attention at the new stage of economic reforms is to be paid to direct foreign investment, having essential privileges as compared to some other forms of economic assistance. Until 1992 and within 1993 the process of creating stable legislative, normative basis of foreign investors functioning will be fulfilled.

In practice it means the following:

> Guaranteeing to foreign investors the property rights on economic participation and formulating the rules of rent for 99 years;
> Guaranteeing property on land when purchasing buildings, constructions or enterprises;
> Adopting and bringing into operation of the laws on mortgage, bankruptcy, strengthening of contract's legislation;
> Introduction of convertibility of the rouble in current and capital operations for residents and non residents, abolishing of limits on repatriation of profits and capital;
> Ratifying international agreements on mutual protection of investment;
> Cutting down the increase of personal profit taxed to 30 per cent.

It is also useful to form in the future the regime for foreign investment in such a way, that the conditions for native and foreign investors in Russia are the same, so called in international practice "national regime", undiscriminative attitude towards foreign investors. The course is to lessen the risk of losing capital, to deduct losses of the first years of functioning from tax payments after the announcement of profits, to establish the order, according to which tax regime, existing at the moment of concluding the contract, will last for the whole period of recoupment. We are going to prove the rights of joint ventures, registered in 1990-1991, on "tax vacations."

Priority spheres of Russian economy for foreign investment's attraction are being outlined at a new stage of economic reform. Such priority spheres are: preservation, transformation, distribution of food products, conversion of defence industry, restoration of fuel and energy sector and creation of extremely important import-substituting industries.

In the next few years attracting foreign investment will be fulfilled in free economic zones. These zones get ex-territorial status in Russia, there will be a privileged regime of external and internal economic activity for foreign and native investors and also all the necessary production and business infrastructure, corresponding to world standards. Zones of such a type will be created in or near the large international seaports, airports, railway stations, if vast territories, labour force and specialists are available.

In the near future, taking into account the complicated financial and economic situation in the country, the main efforts will be concentrated on the creation of some 5 or 6 zones of such a type, for which large financial and material resources are not needed. Among them are Kaliningrad and Sakhalin areas as free economic zones of a complex character.

And still, the main direction will be the attraction and participation of foreign investors in the privatization of Russian enterprises. Contacts are already made with many potential investors, much of which represents large international companies, possessing new technologies and big financial possibilities. Expected amount of privatization projects with foreign investment amounting 50 billion dollars each, according to preliminary estimations, may reach 20 or 30 billions in 1993.

The next stages of the reform are expected to be characterized by mass privatization with wide participation of foreign capital. With this purpose the information system for foreign investors about investment projects of Russia is going to be created, and a number of new legislative norms are going to be adopted, so that to ensure legislative defence and guarantees of foreign investment.

Conversion of the Military Industry Complex in Russia

Alexandre E. Scheindlin, Moscow International Energy Club, Moscow, Russia

For the past few years the world has seen such changes that could hardly be expected.

The Cold War has ended, the dangerous confrontation of East and West has ceased, the Soviet Union has perished. The communist ideology that was stubbornly trying to prove the superiority of the centrally planned economy has suffered defeat. The countries of the former Soviet Union and Eastern Europe have entered the hard period of transition from centrally planned economy to market one.

All these developments occur with the background of huge accumulated amounts of military hardware, ammunition and means of mass distraction. It refers, on the one hand, to the countries of the former Warsaw Pact and, primarily, the countries of the former Soviet Union, on the other hand, to the NATO countries and, primarily, the U.S.A.

Such an accumulation of modern armaments is very dangerous in itself given the unstable political and economical situation in the former Warsaw Pact countries and, mainly, in the former USSR countries. If "hot beds" of national conflicts, both old and new, are also considered, the enormous store of armaments appears extremely dangerous.

It is even more important that a large military industry complex, being the instigator of excess military hardware accumulation and arms race, still exists in previously confronting countries.

The military industry complex of the former Soviet Union comprised the elite part of national R&D potentials and a vast military industry structure that included facilities for the extraction and processing of raw materials, production of armaments, means of delivery and maintenance of large military detachments.

It is right to assume that up to 40% of the country's economic potential was spent for military needs. Citing the Russian Minister of Science, Higher Education and Technological Policy, practically 3/4 of state investment was spent on military R&D till 1985.

It seems clear for everybody what a giant task it is to make a proper use of accumulated arsenals and to convert the existing monsters, i.e. military industry complexes of the two former adversaries, to civil purposes.

These are not simple problems. They should be solved with consideration of many diverse political, economic and social issues arising in the process of conversion.

A number of such issues common to all countries questions are dealt with not only by national governments, but also at international assemblies devoted to the conversion.

Not long ago a representative international conference on conversion was held in Dortmund, some of those present here participated in it. The Conference deliberations led to the so-called "Dortmund Declaration", the main points of which I would like to announce:

Disarmament is an essential precondition to respond to increasing economic, social and financial needs and to reverse trends of environmental degradation at local, national and international levels;

Policies and actions in this regard have to be based on an increased public awareness of the interrelationship of disarmament and conversion of military production. This also applies to the sale and trade of arms as preconditions to successfully monitor the flow of arms;

Conversion policies have also to address the gradual transformation of military forces and their equipment into mobile civil units to prevent and combat natural disasters;

Policies of military conversion should address environmental issues. Specific conversion activities should be based on technology assessment exercises taking into account environmental concerns;

The transformation of military related research and science and technology activities has to be addressed as a central area of conversion policy;

Scientific institutions should be supported worldwide in their efforts to give special attention to the research on economic, technological, sociological, organizational and environmental aspects of conversions and to seek new forms of international research partnership, and in particular, in co-operating with institutions in developing countries and countries with economies in transition;

International development co-operation should recognize the legitimacy of conversion efforts and support related activities in developing countries by providing technical expertise and financing on concessionary terms.

Let us consider the specific situation in the countries of the former USSR related to the conversion problem.

Primarily, I should note the undoubtedly very high level of the military industry of the Soviet Union that ensured the quality of armament which is not inferior, but often, superior to other countries' armaments.

In this connection I would mention nuclear weapons, rockets (missiles), aerospace technology, aviation, armoured, vehicles and submarines.

These achievements are naturally based on a vast corresponding infrastructure in extractive industries, the nuclear industry, energy, transport, chemical technology, metallurgy and communications, etc.

In the background of successful developments there was a network of R&D centres equipped with advanced national technologies in physics, chemistry, mechanics, electronics and other branches of science.

Until recently there were about 1,5 million scientific researchers in the Soviet Union, and a total of 5 million people worked in nearly 5,000 different research centres. In Moscow almost 800 scientific centres had about 1 million people. In Kiev the figures were correspondingly 100 centres and 100,000 employees.

It was mentioned above that a significant part of scientists and technicians served needs of the military industry complex of the USSR. Given the rigidity of science structure it is really hard to make changes satisfying the course of demilitarization in Russia which is undoubtedly an essential problem of conversion. Not long ago President Boris Yeltsin issued an order establishing the national army of Russia. In accordance with the defence doctrine and having no apparent enemies the Russian army of about 1,5 million people will be based on well trained forces of rapid deployment. In 6 or 8 years it might become totally professional, in other words, recruiting will no longer be conscription based.

Thus, entering the period of transition we should have a precise programme of stage-by-stage army reduction including army withdrawal from the former Warsaw Pact countries and countries of the former USSR, which is not easy either.

It will naturally lead to the necessity to liquidate and sell excess armaments.

Full scale conversion of the military industry is even more important for Russia. It is not an easy task if you take into account the dimensions of the military complex in Russia.

Two years have passed since the beginning of the conversion process in the Russian Federation within the framework of the former Soviet Union. However, only after all political decisions on the foundation of the sovereign state of Russia and disintegration of the Soviet Union, on creation of the Russian national army, its quantity and military doctrine are taken, it seems possible to elaborate plans for real conversion in Russia.

One can single out 2 stages of conversion of the defence industry underway in the country for 2 years.

First stage (late 1988-first half of 1991) is characterized by relatively small, determined by "high authorities", reduction of military expenditures, armament and military hardware trade, investments to defence R&D.

The conversion started in 1988 after the agreement to liquidate medium and smaller range missiles, which had a political significance rather than military or economic importance.

Much more important was the decision to reduce within several years military expenditure by 14%, including 19,5% of expenditure reduction for buying armament and military hardware and 15% for military R&D. The indicated reduction was quite noticeable but it did not shake the grounds of the military industry.

These are the average figures for the defence complex. For some types of armament a rather large and painful reduction of production volumes was envisaged. Thus aircraft production was decreased by a factor of 1,8 in 1991 as compared to 1988, production of tanks was reduced by a factor of 2,1, strategic missiles by a factor of 2,4, ammunition by a factor of 2,8, etc. The decrease of military orders and the released material and labour resources were supposed to provide for considerable growth of civil production at defence enterprises. The major task was defined as the correct identification of conversion priorities, formulating proper missions and orders to the defence enterprises. It did not come at once that such an approach would require a substantial state support in the form of centralized capital investments assigned to reorientation of the facilities and engagement of civil production.

To characterize the conversion process a general indication was used, it showed the correlation between military and civil production in enterprises of the defence complex. In 1988-1989 the correlation was 60% to 40%, in 1995 it should become 40% to 60% following the forecast of the state conversion programme.

There figures were to be reached by much desired but not economically sustained growth of civil production.

The figures of the programme should be stated as unattainable even for the period of its consideration in December 1991. The programme did not take into account the process of disintegration of the centrally planned economic system and some specific feature of the conversion process.

The envisaged reduction of arms and military hardware production was not accompanied by a real release of productive capacities because the partial reduction of military production did not allow for the dismantling of the enterprise technological lines and to use the equipment for civil production. Beside, a great portion of the released facilities were temporarily closed forming a mobilization reserve. Thus, the envisaged reduction of tank production made the cost of every item twice as much, 38% of the total expenditures for conversion in 1990 went to compensate the losses the enterprises suffered because of series reduction.

The conversion process and release of resources for civil purposes have been slowed by many factors. Active opposition of the former leaders of the military industry complex was the main reason. They managed to make the investments into military budget reach 96,6 billion roubles (at present price rate) in 1991. Official military expenditures reached the unprecedented level of about 40% of the overall state budget, the typical war time level. At the same time many defence enterprises expected the continuation of the previous military programmes and preservation of the established relations in funding maintenance and management of the defence complex. Some plants still have in store military hardware not redeemed since 1991, but enterprises have already taken credits for large sums of money to cover the needs of 1992.

The period up till the middle of 1991 can be described as the stage of "direct" conversion that was forced by joint efforts of central bodies of the Soviet Union.

The second stage (beginning from the second half of 1991) could be characterized as the stage of "falling" conversion.

The present political and economical situation is fundamentally different. The states of the CIS, including Russia, practically have no budget funds to finance purchases of arms and military hardware, to make capital investments for military complex development and conversion. The sums assigned for the mentioned purposes would be obviously small. In 1992 the budget of the Russian Federation limits arms purchases by 6,5 billion roubles. As compared to 1991 it means an 8-fold reduction of assignation in comparable prices. The science has even worse financing. An 11-fold reduction of science financing could lead to drastic decrease in the number of scientific researchers and technicians to about half a million.

Given the social and political aspects it can be assumed that the main part of military expenditure of the budget will be spent to fund the construction of military personnel apartments, and also to maintain military hardware in the army. Though the present state of the economy makes the solution of the problems only by means of budget revenues highly improbable.
The sovereignty claimed by the former Soviet Republics and their new status of statehood requires radical revision of amounts and structures of military expenditures.

The majority of the former "small" Soviet Republics strive to minimize military efforts limiting them to expenses for maintaining regional formations. To solve the task for coming years the accumulated armament will suffice. Production facilities will be wanted only for maintenance and for limited reproduction.

"Big" republics, primarily Russia, but also the Ukraine, Kazakhstan and, probably, Byelorussia, are entitled to the right to pursue the military policy of the Union and the conversion. It looks like the dominating tendency will be to limit the defence production by the frames of the Russian Federation with possible restricted and voluntary participation of other republics in military orders and contract grounds.

The State Union programme should be transformed into the Russian programme aimed at defence enterprises situated in the Russian territories.

Let us consider what principles should determine the activity of military industry complex in Russia, the principles meaningful for any great country. Figure 1 gives the idea schematically.

Let us start from the fact that the former military industry complex of Russia cannot be totally converted. It should be partially retained to provide the security of the country and in some case it should work for export given the excess production capacities of essentially non-convertible industrial structures.

Thus, upon thorough consideration, the available military industry complex could be divided into two parts: non-convertible and convertible.

Let us consider the missions of the retained non-convertible portion of the military industry complex. Following Figure 1 this portion should provide:

Armament and maintenance of the national army of Russia;
Armament and maintenance of strategic forces of the CIS;
Arms export;
Military technology export.

In accordance with the aforesaid, it seems reasonable to concentrate at our meeting on the promising possibilities of the military industry complex of Russia to export advanced armament and military technologies, some of which would be described in detail in the reports of my Russian colleagues.

Let us switch to the convertible section of the military industry complex. That part could operate upon certain conditions demanding not only political decisions but also investments both from state and commercial structures, as well as sufficient time for effective conversion.

In my opinion the missions of the convertible section of the military industry complex could be as follows:

Civil production for the consumers' market of Russia;
Transfer of new technologies to civil industry;
Civil manufactures export;
Export of technologies of civil needs.

Following the aforesaid let us consider some interesting branches of the military industry complex from the conversion stand point.

Aviation. Achievements of the Russian aviation industry are impressive, they are based on rich network of R&D centres, for example, Zhukovsky Central Aerodynamic Institute, Aircraft Design bureaux named after Tupolev, Il'ushin, Mikoian, Yakovlev and others, aircraft engine centres established by eminent designers, such as Mikulin, Lulka, Kuznetsov and others.

Almost all aircraft design centres engaged in military aircraft construction produce civil aircraft as well. Reallocation of available production capacities often allows an increase in the production of civil aircraft in the course of intense conversion of aviation industry.

In the case of some R&D centres and production capacities it is rather complicated or even unreasonable. It is hardly justified to essentially diminish efficient production of well-known MIG-29 fighters. To trade the fighters for export is easier and mutually beneficial. Effective conversion of military aircraft engine construction could be exercised by using highly reputed types of gas-turbine aircraft engines as gas-turbine power generating units, and driving gas turbines for gas pumping compressing plants on gas pipes. In every case engines need minor alteration by installing a power turbine (plant).

We can already speak about production of efficient gas-turbine units for electric power plants with capacity of 60 MWatts and about mass production of analogous units with capacity of 20 MWatts. The efficiency rate of gas turbine units in gas-steam modification can be over 50% and heating mode can raise the efficiency rate of fuel consumption up to 80-90%.

Aerospace Technology. Conversion allows using at a broader scale communication satellites. Satellites and orbital stations could be used for testing some space technologies, including material and biological developments.

Global and regional monitoring of land and water surfaces of the Earth could present interest to countries with corresponding projects.
The system could include aerial, on land and water stations registering main parameters of the environment and transferring the data via satellites to scientific centres for processing. Such a system would operate satellites equipped with specific devices, satellite launches and ground centre of control, collection and processing of the data.

The environment monitoring programme will pursue the following aims:

To monitor the state of aerial and water basins, land surface and flora;
To produce a proper data bank;
To elaborate recommendations to protect the environment and to maintain sustainable development of regional environment.

In relation to conventional opportunities for space technology a special interest is aroused by many designs and corresponding technological solutions applied in propellant, cryogen and electric rocket engines. Their operational conditions are close to extremes. Thus, the technologies transferred to other branches could be really helpful.

Materials and constructions. Military needs in Russia promoted establishment of major material research centres of general purpose and special research centres for working out most advanced materials and constructions in nuclear, rocket, aviation and other branches of military technologies.

The conversion tasks will arouse interest to materials and constructions enabling to produce articles from different composites including carbon plastic, ceramic and heat resistant composites; metals such as titanium, zirconium, molybdenum, wolfram and the alloys of theses metals, also beryllium and its alloys.

Rare earth materials and their compounds can be efficiently used in civil production. In many cases it is very important to receive super-pure materials, for example, aluminium, etc.

Problems of production of efficient materials for superconductive devices have been solved in Russia. For example, industrial production of superconductive cables has started.

Non-conventional types of military technology (lasers, high energy sources, pulse technology, etc.). Non-conventional military technologies have appeared quite recently thanks to advanced science achievements. Basically, advanced military technologies can always be applied for civil needs.

First of all, let us dwell on lasers, which we can subdivide into high power and low power units. The first ones have more or less destructive effect, the latter category can be used in diagnostics and in control systems. Effective research and industrial complexes established within a short period of time managed to satisfy the required scale of needs in laser systems.

Within the frames of conversion some types of solid state lasers up to capabilities of hundreds of Watts and gas lasers of high capacities of dozens KWatts can be used with high efficiency. All known types of lasers can be used to control and diagnose a great range of civil purpose devices and equipment.

The Russian industry is ready to produce lasers not only for domestic needs but for rather extended export as well, at competitive prices.

The conversion made possible the creation of electronic accelerators, some of which provide for appearance of relativist electronic beams. Such accelerators are applied, for example, to speed up the polymerization of plastic, to process materials. including welding, to purify flue gases of sulphur and nitrogen oxides, etc.

Achievements of pulse energy technology open vast perspectives, for example, shock waves could produce a number of technological effects.

Means of destruction. Conversion of industry engaged in production of ammunition can present possibilities for manufacturing equipment for oil and gas complex of the country, for blasting operations and many other purposes.

Nuclear weapons. Reduction of nuclear arms exercised by planned decrease of nuclear warheads is not an easy act, mainly from the political point of view. We can assume that the process of nuclear disarmament will still continue and the question will rise of utilization of energy potentials of the deposed warheads. In accordance with the available calculations the efficient civil utilization of fission filling of nuclear warheads would provide nuclear power plants with nuclear fuel for several years.

Even nuclear submarines can undergo conversion. At present Russia explores at a large scale oil and gas deposits on Northern seas shelves. The existence of such deposits is proven, but the exploitation is rather complicated because of the Arctic ice situation. The available world practice of shelf deposit exploitation is based on using sea platforms for drilling and primary processing of oil and gas before delivering to consumers.

The ice cap covering the Northern seas practically for the whole year makes using such platforms impossible. It has prompted the leading Russian submarine producing organizations to develop a most interesting technology of extracting shelf oil and gas with the help of properly re-equipped nuclear and conventional submarines. All the required infrastructure for extraction and primary processing of oil and gas would be based on tested technologies and production capacities of the industry producing nuclear submarines. The technology could prove effective not exclusively for sub ice application.

It is obvious that the given examples of conversion of Russian powerful military industry do not encompass all possibilities. Though they prove the prospects of conversion process. It is also obvious that the conversion would require proper investments and reasonable time for implementation. We presume that the changed political climate would make this huge endeavour beneficial, provided close international co-operation.

144

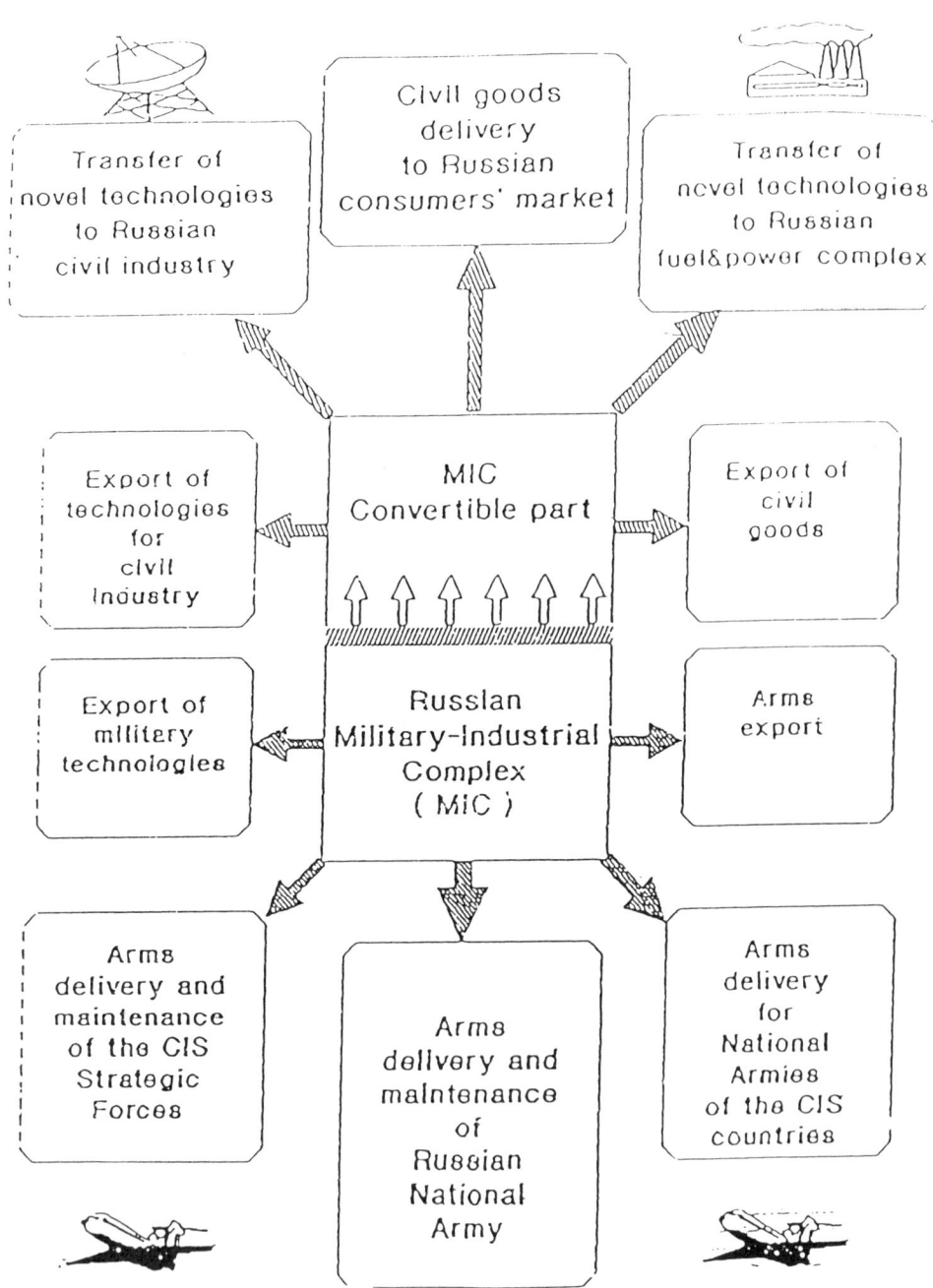

Figure 1